Bee-Keeping

With Twenty Hives

A.L. Sandeman-Allen

Travis & Emery

Arthur Leonard Sandeman-Allen:
Bee-keeping with Twenty Hives.

First published by Bee Craft in 1952.

Republished Travis & Emery 2008.

Published by
Travis & Emery Music Bookshop
17 Cecil Court, London, WC2N 4EZ, United Kingdom.
020 7240 2129
neworders@travis-and-emery.com

Hardback: ISBN10: 1-904331-71-8 ISBN13: 978-1904331-71-1
Paperback: ISBN10: 1-904331-72-6 ISBN13: 978-1904331-72-8

To

MY WIFE

and

to those who helped me with this
book and with my bees, especially
Sir Edward Travis, H. Pagan,
C. R. S. Richardson, F. J. Miller,
and my Father.

CONTENTS

ACKNOWLEDGEMENTS

I am indebted to The Fruit Grower and The Bee World for permission to make use of certain material which has already appeared in those magazines, and to Mr. C. H. Hooper F.L.S., and the R.S.A. for permission to reproduce the plate of pollinating insects.

FOREWORD

It seems so often with books on bee-keeping, that the authors know so much more about writing books than they know of practical bee-keeping and it is therefore particularly refreshing to read a book such as Mr. Sandeman Allen's which is not just a re-hash of earlier books on the subject, but gives us his own experiences and difficulties and the ways in which he has overcome them. Moreover, he fills a gap in writing particularly for a large section of the bee-keeping community whose special needs and problems have not hitherto received the attention they deserve.

As a bee-keeper myself, with my own ideas, I do not see eye to eye with Mr. Sandeman Allen on every point of management as is only to be expected, but on the whole I believe that not only the special class for whom he writes, but bee-keepers as a whole, will find it well worth while to read this book and few will fail to find information therein which is of much value.

E. W. D. MADOC,
White Hall,
Saham Toney,
Norfolk.

CHAPTER I

INTRODUCTION

THIS book is intended to be a practical guide for the increasing number of bee-keepers who wish to keep about a score of hives for honey production as a spare time venture. It is not for the Commercial Honey Producer, whose problems are of a different character, and it is certainly not for that group of people who are usually self-described as " beginners " and who never seem to be inclined to get past that stage. I shall attempt to describe a system of management which I think is suitable for those who have a few hours a week to spare and wish their hobby to render them a modest but worth-while profit.

There are books on bee-keeping by amateurs, and the authors, consciously or unconsciously, tend to bring into their writings the stamp of their professions or occupations. I think this makes their books all the more interesting. I am an Accountant and I know I tend to look upon bee-keeping in a manner more practical and " costed " than romantic.

As this book is a practical guide to management, I have not included any chapters on anatomy or on disease *per se*. A knowledge of anatomy I regard as nearly, but not entirely, useless to an intending honey producer, the main exception being that it is desirable to be able to dissect bees to diagnose the presence or absence of acarine and nosema. Disease is a subject which requires a specialist, and by omitting it I hope it will not be thought that it is unimportant. It is very important indeed.

My aim is to get the maximum amount of honey in the time I have to spare, at the smallest cost, and (since I do

7

not make a living from bees) in a way that will give me the most enjoyment. The last part of that sentence occasionally conflicts with the first, but I do not care. I work at a desk all the week, I enjoy my weekends with my bees, and I intend to continue to enjoy them. I do not mean this in a sloppy way—that I live for my weekends with my (beloved) bees or anything of the kind. My work is interesting and I hope useful—but a complete change at weekends does me and my work a power of good.

As I intend this book to be brief I have left out many chapters which are usually found in guide books. There are almost too many books today which explain to " beginners " how to take a swarm or how a feeder works. More important to my mind is how best to use that swarm and when to use that feeder. So I assume my readers are already bee-keepers and have already read books telling them how to " begin." This leaves me free to discuss those special problems which arise for those who want to keep a fair number of hives, and can only spare weekends and perhaps an odd half hour in the evening between arriving home from work and the evening meal.

My bees are fairly yellow. My hives are Modified Dadants. I think my efforts to manage the one in the other have met with a fair degree of success. The apiary is near the coast in East Anglia, on the boulder clay belt. This clay seems to take a lot of warming through before plants grown on it yield much, but when there is a honey flow, it can be terrific.

Readers will appreciate that what may be a good system of management for yellow bees, large hives, cold winters, rainless summers, flows early and late, etc. etc., may be unsuitable for black bees in Scotland or grey bees in Devon. I have been told by a commercial bee-keeper that the strain of bee counts for more in the scheme of management than the size of hive and I am inclined to agree with him. No doubt I could write a far more comprehensive book if I had had experience of managing more sorts of bees in more sorts of hives in more parts of England, but then I would not be a weekend bee-keeper and so well aware of his peculiar problems.

CHAPTER II

HIVES and FRAMES

I FOUND the following in a recent book on bee-keeping where it was in heavy type :—

" The Beginner must always remember that hives are
" merely boxes to house bees, with varying degrees of
" comfort and suitability . . . . ."

I cannot recall finding in any other book, say on cows or fishes, anything like " The beginner (!) must always remember that cowsheds/aquaria are merely buildings/tanks to house cows/fishes, with, etc. . ."

Both the above sentences are in one sense truisms and in another sense bunk.

Bees can get along quite well in boxes of almost any shape or size above a certain limit. The problem is to find a box and frames which will enable the bee-keeper to obtain the largest possible crop with the least labour and cost. If the box is unsuitable for the bees there will not be much crop ; if it is unsatisfactory for the bee-keeper he will not be able to manage sufficient hives to get a profitable crop in the time available, and if it is too expensive it will not be worth keeping bees at all. Between these extremes are all the well-known makes of hives. There are plenty of makes to choose from or you can have them made to your own design if you like, but it is important to remember that on the size and type of hive to a very great extent depends the whole system of management.

Most people thinking of keeping a score of hives already possess a few, and also have fairly decided views on what sort of hive and frame is best for them, but in spite of this I do not think it is a waste of time to debate the subject.

Broadly speaking, hives fall into three main classes ; " W.B.C. type," single broodchamber hives and double broodchamber hives.

With one exception, all the recent books I have read condemn the W.B.C. hive, and I was beginning to think that this old stager had seen the end of his useful life and was being slowly and very respectfully placed on the retired list. It surprised me to find that it is still the best seller. Among

the questions circulated to the members of the Association of
which I am Secretary was : " If you were starting afresh,
what hive would choose?" As might be expected there was
not very many answers. The National hive was a clear
favourite but still one member in five preferred the W.B.C.
Only one member preferred National Majors, and nobody
mentioned Langstroths or Smiths. One brave lady wished she
had M.D's. I have studied these returned questionnaires and
I have talked to many people but I cannot find out why the
W.B.C. is such a good seller.

I manage a few W.B.C's for a friend and I regret to say
he never gets much honey. Maybe I have not got the right
" touch " for the things. As for the lightness of the supers,
his are seldom much trouble even for me. As a result of a
motor-cycle accident I have a slightly damaged back, so I
have to hire help when my own supers are taken in to be
extracted, but even so I prefer them to be heavy.

I know I do not write with a great many years of experience, but I manage a few and have helped with many,
and I have come to the conclusion that it would take me
about one and a quarter times as many hours of work to produce half a ton of honey from W.B.C. hives as from
Dadants. To my way of thinking that alone rules these relics
out of all serious consideration. In addition the initial cost
of hives required and equipped to produce an average of half
a ton of honey a year is greater if those hives are W.B.C.
There are no end of other troubles. Mice are more difficult to
keep out, so are wasps and various creatures which thrive
in between the double walls. If that is not enough, W.B.C.
hives cannot be moved in the active season when full of bees.
I know there are published photographs of the thing sup-
posedly being done, but I maintain that those hives are only
half full of bees and you cannot possibly tell that from a
photograph.

Most people seem to agree that W.B.C. hives are more
decorative. From now on I shall ignore the species.

As I have said before, most people who are thinking of
keeping a score of hives already keep bees, and are probably
thinking that as they already have bees on British Standard
frames it is best to choose a hive which " takes " B.S.
frames in order to comply with the commandment to be

found in nearly every book—" Aim to have one type of hive and one size of frame in the apiary. An assorted collection is wearing on both energy and temper." It is quite a good rule as rules go but is certainly not golden. Frequently the advocates of the rule also fancy some divided or other frame for mating hives. I keep half a dozen Nationals and they are most useful indeed. They contain B.S. frames with their long lugs which are, I think, far and away the best frames for use in nuclei. My Dadants are too big for the job and as for shallows, I would want to put them up on end to make a *section de chauffe*, but B.S. frames are perfect. Besides, people sometimes want to buy my bees and they usually ask for B.S. frames. I keep my breeder queens in Nationals for convenience, and for other reasons given later, and I put all my nuclei into National bodies in the autumn and pile them up on top of each other, uniting with newspaper. After a good feed they go into the winter as strong stocks. Next spring some make nuclei, some are sold and some make sections. Although I do not use them for the production of extracted honey, I would not be without a few Nationals. If I had not started with Nationals and naturally do not want to scrap them, I would use those hives Manley describes which hold 12 frames and are of the right size to take M.D. supers.

So, if you intend to increase by over a dozen hives and already have some B.S. equipment, keep it, it will be useful—and give some thought to the other hive types without fussing too much about " One Apiary, One Hive and One Frame."

The National Hive is not at all a bad hive but it is best to work it with two broodchambers or, as more generally described, double broodchambers. If the bees are to make the best of a honey flow there need to be plenty of them and if the queen is restricted to a small laying space she cannot lay enough eggs to produce a strong colony. In a double broodchamber she has plenty of room. Wadey gets excellent results from using doubled Langstroth hives which give an enormous brood nest, so there is no need to worry about giving too much room. An ordinary Dadant broodchamber is very little smaller than a double National. Single broodchamber Nationals will give good honey crops, sometimes

year after year, their owners say. No doubt they do in my
district, but if they beat my average their owners have not
told me about it. Also a single small broodchamber requires
a strain of bee of a sort which is not very prolific and is long-
lived, the latter quality being rare and difficult to spot and
to fix in a strain.

My own feeling is that it is simpler to manage bees in a
double broodchamber than in a single. There are countless
manipulations which can be carried out for which a double
broodchamber is either essential or it makes the job more
simple. " Snelgroveing " is an example of the first, and such
things as aggregating the brood or removal of unsealed brood
are easier to do with many small frames than a few large
ones. Another advantage is the shape. Bees seem to do better
in a shape like [] rather than like ⸗, and it is obvious that
it is more economical in bees to keep up the temperature in
a broodchamber of the first shape. This is clearly demon-
strated in the diagrams in Wedmore's book *The Ventilation
of Beehives* in which he shows the cluster covering a few
frames deeply in cold weather. Bees do not expand horizontally in the spring as fast as vertically, and if they do
expand horizontally too fast the outside frames may be chilled
if there is a cold snap. In fact there is quite a formidable
list of advantages and many a long list has been published
in favour of the double broodchamber scheme, but I think
in one way these lists miss the point. I give two hypotheses
which I believe to be true. Assuming the same total size of
broodchamber, i.e. one single large box as against two
smaller ones.

1. Routine operations take longer to perform in double
   broodchamber hives.
2. A little more honey may be expected.

Well is it worth it? In the time available can one get more
honey in total from double than single broodchamber hives?
I should think there is very little difference indeed, but there
are however other factors. The doubled National costs more
than the single Dadant. The National extractor costs quite a
lot less. The double National is more of a job to transport
and is nearly impossible for one man to lift about when full.
There is an irritating point with Nationals ; either you suffer
gladly or sadly the presence of metal ends or else you use

Hoffman B.S. frames which are a very tight fit at twelve to the broodchamber. The Dadant is more sturdily constructed, lasts longer, and stands up to rougher treatment. Finally, I think bees winter better in double broodchambers. Incidentally, with regard to point 1. above, routine operations in colonies not attempting to swarm can be very speedy if the top chamber is tipped so that any queen cells can be seen along the bottom edge of the frames in the top box. You will see the presence or absence of cells in almost every case, but you will never see the presence or absence of foul brood that way. It is a common practice. It is a most dangerous practice in the hands of amateurs.

I have weighed up the pros and cons and on the whole I am glad I have Dadants mostly and a few Nationals. If 1 started afresh with twenty hives, I should have seventeen Dadants and three doubled " Nationals " with twelve frames in each so that floors, roofs, excluders and supers would all be Dadant.

- So far I have compared a double National with a Dadant, and this comparison is really between two broodchambers of nearly the same size, although the Dadant is slightly smaller, but in any case the result is a large brood body. A single broodchamber National I would describe as a small brood body—too small in fact except for a special strain of bee. There are, however, two hives having medium sized brood bodies, and these are the 16 × 10 Modified Commercial hive, and the Langstroth, and these roughly correspond to one and a half National bodies. The latter is more like ▭ in shape and the former rather more square. I have had no experience with Langstroth, but owing to its shape I would prefer to use it doubled. A double Langstroth would, however, be about the most expensive hive one could choose. Many people would say it is far too large ; it may be but I am not sure and again it depends on the strain of bee. It has often been said that such and such a hive is " too large " and this has been said about Dadants and any double brood chamber hives, in fact Herrod Hempsall devoted a line in a list of " Don'ts for Beginners " crusading against any hive with *more* than ten B.S. frames. But I have yet to find a clear and logical argument saying just why any hives are too big. As far as I can see the only argument is that with

a tiny broodchamber the bees have no room to store any
honey during the active season as well as their brood, so if
there is a short sharp flow the honey can only go into the
supers, in fact with a tiny broodchamber you get the honey
sooner. The old orthodox British teaching was then that
you " took " this honey, extracted the unsealed first and
put it back somehow, and then the bee-keeper got together
a few jars to fill with this season's honey. In a year like
1948 such games may work. When the rains came that year
many hives took the honey back from the supers and what
happened to those which had had their supers removed I
would not know. A few I know of became very weak and
a lot of hives in this district died out in the winter of 1948/49
quite likely for that reason.

I like the Modified Commercial very much indeed, but I
think it is just a trifle on the small side, a remark I must at
once qualify by adding—in my district with my type of bee
and with my type of management. It is a good hive, cheap,
sturdy, and very very much better than any other single
broodchamber hive of smaller dimensions.

You can design your own hive, and your own frames too
if you wish. I know a man whose " Nationals " hold ten
frames and another whose " Nationals " hold thirteen. If
you are planning to keep a hundred or so hives it may
possibly be a good idea, but for the week-end bee-keeper the
best advice is to stick to the standard stuff. Even if you make
minor alterations it means nothing can be bought in a hurry
as reach-me-downs, and it is surprising how often we all have
to write urgent letters to appliance dealers for some odd item
of equipment. It is often claimed that Nationals are improved
by nailing them up in such a way that the bee space is
above the frames instead of below. It is in any case a very
doubtful advantage and it involves mounted or Waldron
excluders and different crownboards, but even if the advantage is real, if only a score or so of hives are being kept it
is not worth altering standard equipment.

I do not know whether the bee-keepers in this district are
a fair cross section of the bee-keeping fraternity or not, but
of those I have met who are expanding their apiaries and
still remaining week-end bee-keepers all have B.S. frames
and nearly all have Nationals. If the plan is never to in-

crease beyond an additional ten hives, then it is best to stick to B.S. equipment. A few Modified Commercials for example can be an expensive nuisance. It means buying a new extractor and finding someone to buy the old one. There are, however, exceptions. Supers which can be used on Nationals can be used on Modified Commercials and vice versa. Do not worry unduly about bee spaces because a Waldron excluder will attend to that. The 16 x 6 super frame of the Modified Commercials is a most excellent frame. If the extractor is due to be replaced by a radial, and if there are not already too many National supers, it is probably worth while changing all supers to 16 x 6 frames. As for selling the old drawn combs, just get in touch with the local foul-brood inspector and find out if you can whose supers have recently been burnt! The empty supers are not very expensive in the flat, but the frames are both expensive and troublesome to assemble. Supers can easily be made to hold B/S shallow frames and which are the same superficial area as M.D. hives, so a cheap way of changing to M.D. broodchambers is to have these special supers made. Put the bees into M.D. broodchambers and super them with B/S shallows which will run across the frames in the broodchamber. The National or W.B.C. broodchambers can then be divided and used as nuclei so nothing is " lost " on the change over except the empty super boxes and even these have a variety of uses. There are two big advantages in this method, firstly, the extractor you have can be retained and, if you are starting afresh, the new extractor will be far cheaper than one for large frames, and secondly, (ladies and born-tireds please note) you have all the advantages of Dadant brood chambers with light(-er) supers.

Supers.

The subject is dealt with in every text book, but as individual authors usually prefer the type of super they themselves use, they seem to write with considerable bias. The opinions of the experts differ considerably, so I have attempted to collect together the main points for and against each type of super.

Leaving aside sections, the main points to consider when deciding what sort of super to use are :—

    (a) Whether to have full-depth or shallow frames ;

    (b) Whether to adopt so-called "drone" or "worker" spacing ;

    (c) Whether to use drone-base or worker-base foundation, and

    (d) What shape frame to use and what method of spacing it.

(a) Full-depth or Shallow Frames

One point of view is expressed by Manley who says that a super frame is for honey storage and should be designed for the purpose and not be just another brood frame. There is no doubt that a shallow frame designed for use as a super frame is easier to handle and uncap. Frames deeper than Langstroth deeps are extremely unwieldly in supers and in the extractor.

The exponents of deep frames argue that it is so convenient to have brood frames built out right down to the bottom bar with all worker cells. The best combs are undoubtedly those built above the brood chamber. Wadey goes so far as to say that shallow frames are for " ladies, and elderly, delicate or born-tired men." Even so, full-depth supers full of honey can be too much for many of us if working single handed through a lot of hives, and we do not all want advancing years made obvious by having to buy a complete new set of supers when we qualify for pensions.

It has been given as a disadvantage of deep frames that they lead to extracting from combs that have been bred in, and this in its turn contributes to the spread of American foul brood. I can see no reason why super frames cannot be kept separate even if they are the same size as brood frames.

A satisfactory compromise is to use only sufficient full-depth supers to provide enough combs to replace those which are rejected from brood nests annually, and also for frames for sale in nuclei and for spares.

(b) " Drone " or " Worker " Spacing

The widest spacing in general use is 2 inches from centre to centre (wide metal ends) and the narrowest $1\frac{3}{8}$ inches from centre to centre (Langstroth). For extracting frames it is convenient to have cappings stand out from the top and bottom bars sufficiently to get the uncapping knife under them. The wider the spacing the fewer the frames to pay for and assemble, fewer combs for the bees to seal, and fewer for the bee-keeper to uncap and extract. There are certain disadvantages in using spacing that is very wide which Manley gives in his book *Bee-keeping in Britain*, but Madoc, another large scale honey producer, uses a $1\frac{7}{8}$ inches spacing and many other people find wide metal ends quite satisfactory.

Narrow spacing is better if the bee-keeper is using the " one-and-a-half principle," i.e. a full depth chamber plus a shallow chamber to make up the brood nest. The shallow frames can then be aligned with the deep frames. Of course these "shallow frames " are not " super frames " as they form part of the brood chamber, but if a different pattern shallow frame is used in the supers it means yet another sort of frame in the apiary. If you want a broodchamber only slightly larger than a single National chamber, a British Commercial or a Langstroth is a more simple proposition.

(c) Drone- or Worker-base Foundation

This is another subject which has been laboured for many years. The use of drone base foundation means virtually no pollen will be stored in extracting frames. The honey flows out of the larger cells more easily and this may be quite a consideration with hand-powered radials or with thick honey. It is said that the queen spends much time walking about upside down on the queen excluder trying to get at the drone cells ; I do not know whether anyone has managed to prove this ! If the queen does get through the excluder, you have one dud stock to deal with. It is a rare occurrence, and even if it happens it is only one dud stock and is not the major disaster which some writers envisage. The theory about economy of wax in drawing foundation appears to have been discredited.

The advantages of worker foundation are that the comb can be used for breeding in if required, should there be a faulty excluder, or if no excluder is used. The exponents of worker-base foundation assert that honey is more readily stored in worker cells than in drone cells, and the examination of wild colonies tends to support this view.

This problem of drone- v. worker-base foundation is often confused with " drone " v. " worker " spacing. There is no reason why drone-base foundation should not be used with "worker " spacing (although this would be rather pointless) nor why worker-base foundation should not be used with " drone " spacing.

(d) The Shape of the Frame and the Method of Spacing

There is now on the market a frame generally known as the " Manley-type. frame," and it is made in Modified Dadant, British Commercial (complementary to the 16 × 10 frame) and B.S. size. It is " close ended," that is to say the side-bars are about 1⅜ inches all the way down, and it has been designed for honey storage and for no other purpose.

The Hoffman frames have narrow spacing, which is very narrow in Langstroth and British Standard frames, but they are satisfactory and do not swing if the bees have to be moved.

Metal ends have been described by countless writers as " a nuisance at extracting time." They certainly are. If the hives are not to be moved about, there is really no need for any spacers and the frames can be adjusted by eye quite easily. Yorkshire spacers are better but the spacing is narrow. They do prevent swinging however, and do not have to be removed for extracting.

" Teswain " spacers are obtainable in wide or narrow spacing for 10 or 11 frame British Standard hives, and work very satisfactorily in supers although they are an abomination in brood chambers.

The width of the top and bottom bars is a final point for consideration. The wide top bars now obtainable for British Standard frames are a little stronger, and are good for use with spacing which is wide enough for the cappings to stand out from the wood. There is also more wood for the bees to attach the top row of cells to. There is something to be said for having the bottom bar the same width as the top bar, so that the uncapping knife can be run along in contact with them both.

While I am on the subject of frames, there seems to be a great drive in favour of top bars which are nearly square in cross section with a device called a " heavy wedge " to hold the foundation. With B.S. frames I think this is nonsense. With larger frames the need for a sturdy top bar becomes more pressing. I have had some difficulty in getting Dadant brood frames with top bars ½in. deep, not weakened by cutting out a " heavy wedge " and containing a groove down the middle in which to fasten the foundation with melted wax. A Director of an appliance firm told me these top bars would sag, so I put nine in a super and each Dadant frame held about 10lbs of honey. There was no sagging anywhere. In any case it is easier to put foundation into a groove than to fiddle about with wedges, and still easier to re-equip an old frame after melting.

Nucleus Hives

There are three main ideas on nucleus hives. A National hive can be used either divided into two or three parts or just restricted with a division board. A special hive can be used. A travelling box can be used. Of course a proper National is more expensive, but it does have the advantage that it can be used for its normal purpose if and when required, and in the autumn can be made up to eleven frames and wintered either by itself or united to an existing hive. Travelling boxes are a good idea. Quilts or inner covers of three ply, with cleated edges, can be used, and when required as travelling boxes the inner covers are removed and the ventilated covers substituted. It is a good plan to make them with plenty of room under the frames. Special nucleus hives can be obtained from all appliance-makers, and these are very nice to use, but with few exceptions cannot be pressed into service either as hives or as travelling boxes. A six frame travelling box makes an excellent mating hive, and either one or two frames can be removed and a frame feeder substituted if required.

Roofs

A word about roofs. Deep roofs cost an appreciable amount more than shallow ones and have certain well known advantages and others which are not so well known. The usually quoted one is that there is no need to put a brick on top to stop the lid blowing away. This is surely a very small advantage and is only true if the roofs are loose fitting. They do stop the woodpeckers getting at the hand-holds. Rain beating down in winter only strikes the lower part of the brood box and the whole side does not get glistening wet and, when the wind blows on the wet area, probably the hives with shallow roofs are colder. On the other hand the sun cannot strike the whole side, warm the hive up a bit, and give the bees a chance of a cleansing flight. When only one super is on, a deep roof will cover the join between super and brood body. A strip of wood can be nailed round the inside of the roof so the roof can actually rest on the hive on this strip which leaves a few inches between the crownboard and the roof, enough room for one of those flat, round feeders. My own hives have shallow roofs except for two, one of which warped and so it is shallow now. I would rather have deep roofs if they were the same price, but they are not. I doubt if there is much advantage in them.

Floorboards

You can buy supers and broodchambers from different firms and it makes very little difference. A mixed lot of roofs is a little nuisance, but not much. However, a collection of different floorboards can be really irritating because you will find different sized entrance blocks are made by different manufacturers, and these are not by any means always interchangeable. If you can afford it, buy a few extra floors and still more entrance blocks when you begin, and trouble will not come upon you so soon.

Some floors, notably the standard National, are reversible. The Dadant floors are not, and if I was starting again, I should have reversible Dadant floors, but now I must wait until they wear out, by which time there may be some better pattern.

An advantage of a reversible floor is that when the floors are cleaned in the spring, the floor may be examined and put back the other way up. The usual method is to use a spare floor, or to clean the existing one while the broodchamber is on one side. It is easier to use a spare floor, but *there is some evidence that moving floors from one hive to another may contribute to the spread of foul brood.* .

I wish a floor could be contrived whereby a thin sheet of some metal or composition substance could be slid out at the front or the back of the hive, cleaned or reversed and slid back. The floors could be permanently stapled on or the broodchambers would be made to include the floor, and the spring cleaning would no longer be a three-man job or a difficult job for one man.

CHAPTER III

ANCILLIARY EQUIPMENT

HONEY extractors, wax extractors, cappings melters, uncapping knives, ripeners and strainers will be dealt with in chapter 7. Hives have already been given a chapter, so there only remains the smaller items.

Feeders

Autumn feeding is immeasurably more important than Spring feeding, and if a different pattern feeder is more desirable for Spring it is just too bad unless you are prepared to lay out more money. Just a few deep-lid type 7-lb honey tins with holes punched in the lid are useful for Spring work however, if you have a few spare empty supers to accommodate them. Apart from this, considerations for Autumn feeding should govern the choice of feeders. Once syrup is made, either it must go straight into feeders on hives, or it must be stored until required. This rather obvious remark means that the weekend bee-keeper must either make his syrup and feed it during the week, or he must have sufficient feeders to get the whole job done in three or four weekends. This narrows the choice of feeders to one very large one per three hives or one smallish one per hive. If you have big feeders on one third of the hives, you fill them one weekend, and by the next weekend those hives will have been fed and that is that ; the feeders are moved to the next hives. If small ones are used, each feeder is refilled two or three times before the job is done. I get along very well by having one big Miller feeder per three hives, and one filling is nearly always enough for any one hive. I lift the roof, put on the feeder, pour in the syrup, replace the roof and *provided the roof is bee-tight*, that hive should be set for the winter. There may be a few hives that require a second helping, but those can be left until the end. Any other type of top feeder requires either an empty super, an empty roughly made box like a super, or a deep roof with an arrangement to hold it clear off the crownboard, or else the whole weight of the roof and the feeder will be taken by the middle of the crownboard, which cannot do the crownboard any lasting good.

Frame Feeders

These are very useful in nuclei. Bees will empty a frame feeder far quicker than any other sort, and even quite a wide frame feeder, if filled in the evening, is empty by morning. It is as well to have a few of these for use when robbers are active ; fill them in the evening and you will find all the " fuss " over before robbers start out next morning.

Smokers

This subject has been debated *ad nauseam*. Practically every modern author uses no end of space proving that those little upright smokers are useless relics of bygone days. I think like nearly everyone else that a big smoker is a real necessity. An 8 × 4 inch size, in copper, will last years and copper is worth the extra cost. I have met many who have used both sorts and prefer the big bent nosed type, but never yet one who has given a proper trial to both and still prefers the upright kind.

For the man with twenty or so hives, however, I suggest both sorts. The extra little one is only a matter of a few shillings and is so useful. If there is a difficult stock to go through one evening, or a couple of nuclei requiring mid-week attention, it is so much simpler to light up a roll of corrugated paper for the upright than to get out the big one and go to the trouble of getting it going for just four puffs. Also, the little one is the sort required if you use the sulphur treatment against acarine.

Old sacking is *the* fuel. If you do not know where to get
it, the local corn merchant will tell you. When you've finished
for the day, push a tuft of grass down the spout and the
smoker will go out. Next time, the half burnt roll of sacking
will light up easily. *Never* use sacks which have contained
artificial manure as the honey may be spoiled.

Hive Tool

Two (or three) of these are needed and the shinier the
better. If you are terribly tidy-minded, one will do. I have
lost mine so often that I carry two ; there just is not the
time to go hunting for a missing hive tool which usually
turns up later in the afternoon.

Clothing

A white boiler suit with a zip up the front and spud-net
underwear is just the thing. A cotton shirt and a pair of drill
trousers, both old (but darned) is nearly as effective. On
days when the bees may be expected to be temperamental,
a pair of rubber boots will stop them marching up one's legs.
On other days, either you wear a pair of cycle clips or keep
cooler without and get an occasional sting. I am informed
that although bees run up inside trouser legs, they will not
normally run up inside shorts. Perhaps it is because these
crawlers start at ankle level and fly off before reaching knee
level when there are no trouser legs to stop them. Perhaps
they are just good sports. Gauntlets can be made from linen,
old sheets are excellent—with elastic at wrist and elbow, but
make them long enough so the top elastic does reach above
the elbow. If gauntlets are not worn, bees will get in at the
apertures above the cuffs ; roll up the sleeves and you will
get stung less—perhaps for the same reason as the wearers
of shorts. Regarding gloves, it is no good being dogmatic.
If you *can* do without gloves, then do without them, but
have a pair just the same. If you wish to clip a queen, it
is far easier to do it with clean fingers and that means wearing gloves until the last moment or there is sure to be some
sticky propolis on them. Also, there are plenty of days when
either you have a skin like a tortoise, or you wear gloves,
or you leave the bees alone, but for the weekend bee-keeper
this last is not always possible because a couple of thundery

weekends will mean a twenty-one day gap between inspections. It used to be fashionable to regard those who wore gloves as being " bad " bee-keepers. This is a narrow outlook. If you are a dentist, a dressmaker or a follower of quite a number of professions or trades, a badly stung hand may mean no work can be done the following day. Propolis will stain fingers and Manley advocates gloves for that reason, but if you keep less than forty hives, a good scrubbing will remove what little staining you will get.

Hive Carriers

These can either be bought or made up by the local blacksmith. I used to think they were not necessary until one day the lorry got stuck in the mud and each hive had to be carried a couple of hunderd yards down a muddy lane. The only points to note are to make sure the hooks are big enough to fit under the side runners of the floors and that the shanks are long enough so the handles are above the point of balance.

Super Clearers

It is handy to have a collection of these with good thick battens round the edges. Either they have to be bought, and they are not cheap, or else made, and that takes time but is very easy. Provided the roofs are beetight, the crownboard can be used. As to the actual escapes, there are several patterns on the market. For some reason or other, the slots in the clearer boards are usually made just too small for a Porter escape. Porter escapes are also made in two sizes, one only a tiny bit larger than the other. The slots are usually made to hold the escape as a tight or " push " fit, so they will not fall out when the clearer boards are carried about. I prefer larger slots so there is no trouble in fitting the escapes. The slots can be enlarged but there are sure to be a few which have been forgotten, and bees can be short-tempered when clearer boards go on. A push fit is probably best for clearer-boards proper, but not crown-boards. There in an escape with ten or so fingers which operate downwards. These are efficient, but if you are using crownboards for clearers, order them with thick battens round the top side (in fact, it is a good plan to do this anyway) because these clearers need space.

Swarm Box

This is only mentioned because you *will* need one. Make a lid with a rim to fit over the box rather than a flat one with a hinge. I keep the lid on with string and it is quite satisfactory. Choose a box on the large side and the more perforated zinc or wire mesh you can incorporate round the sides without weakening the structure, the better. Arrange for a flight hole.

Apiary " Store "

It is quite easy to pick up a couple of old W.B.C. or similar hives, and they can be most useful, provided they are waterproof. The smokers, hive tools, and a supply of sacking can be kept in one, and all manner of things such as queen candy, introducing cages, nursery cages, match boxes, escapes, a few spare frames with foundation, some perforated zinc, etc. can be kept in the other. This prevents a lot of trips back to the house for forgotten oddments.

Frame Holders

These are gadgets to clip on the side of the hive or to stand on their own feet to receive the first frame removed from the hive. It is really simpler to stand this frame on end and lean it against the side of the hive. If it contains brood, it is better to put it out of the wind in a nucleus hive or some rough box.

Dummy or Division Boards

Appliance dealers' catalogues seem not to differentiate between these two things. The difference is that a dummy board has a bee space between it and the hive wall, i.e. it is the same size as a frame, while a division board is larger and touches the hive wall. A division board is used to reduce the size of the hive from eleven to (say) six frames, or to divide the hive into two nuclei. A dummy can be put in at the side of the end frame to fill the gap (if any) between it and the hive wall. If Hoffman frames are used, and are a tight fit, it is better to put a dummy in instead of the eleventh or twelfth frame, making sure there is just a bee space between the dummy and the hive wall—a point that has some advantage in the production of sections. By nailing little bits of wood on to one side of the dummy to keep the face of the dummy clear of the hive wall, swinging in transit can be avoided and fewer bees will be crushed. A dummy board of half or less the normal depth can be used to fix cell cups to.

Hive Numbers

Anyone can devise his own method of numbering hives. The necessities are that the numbers shall not blow away and shall be easily changed round—so don't get those expensive brass or composition things that are designed to be screwed on permanently. I know of four simple systems.

1. Sheet metal squares painted white all over with red or black numbers and having slots like an "8" so they can be fixed over nails driven nearly home on the fronts of the brood bodies.

2. The "lids" from tins of soup, vegetables, peaches, etc., removed with a wheel type tin opener can be similarly painted, and, being small and light, can be put on with drawing pins.

3. Paint the numbers direct on to the fronts of the hive roofs.

4. Use chalk in the summer, and in winter keep a map of the apiary, or numbered pieces of cardboard put on top of the crownboards.

Wiring Boards and Wiring

It is worth while making jigs for assembling frames. Two angle brackets will do, fixed like _| |_ so that a frame just fits upright between them. Cold water glue is a good way of fixing the parts together, and when glued, stand the frame between the angle brackets and drive a nail vertically through the top bar into the side bar to make sure of the job. Side bars can be bought with holes already drilled and this saves quite a lot of time, otherwise side bars can be stacked in a form or held in a vice so that half a dozen can be drilled at once with a long twist drill. Regarding wiring, a " board " to do this can be made quite easily. The principle is to have the frame held steady while the wires are pulled taut and nailed. My own is a length of planking with a " T " piece of 2″ × 1″ along the top, squares of wood in appropriate places to hold a Dadant frame steady and a similar arrangement on the reverse side for B/S frames. Embedding boards can be bought, or you can just saw off a length of plank the right size to fit comfortably inside a frame—which is all there is to it. Embedding tools can be either wheeled devices heated in water, or electrical which people who know agree are better. Some are complicated contraptions with bell-pushes to make the contact affixed, although the switch is far easier to work with the foot. My own arrangement consists of two handles, with one point (an oval brad) on one and two on the other, side by side, only one of which is connected to the battery. To operate, just place the two " live " points on the wire until the wire is in and then take the pressure on the " dead " point for the second it takes to cool. Do not attempt to embed a whole length of wire at once ; it is best to place the electrodes one at the end of the wire and one half way along or even less for a big frame. Too much current will make the wire expand and rise up from the wax. This arrangement caters for all

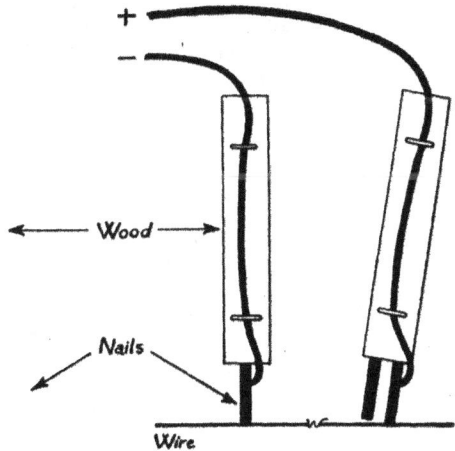

Wire Embedding Device

1 First position. Contact not made.

2 Make contact until wire is embedded. Then return to first position until wire is cool.

sizes of frames. As for the battery, buy one which is waiting
for salvage from the local garage. Each time you need to
use it, swap it for the best of the bunch awaiting salvage,
most garages can find a battery which will take enough
charge to do a hundred or even two hundred frames.

Excluders

It seems agreed by most people nowadays that the type of
excluder used makes no difference to the honey crop. If the
bee space is below the frames, the long slot rests directly on
the frame tops and is satisfactory. If the beespace is above,
the excluder must be mounted and the long slot, when mounted, lets queens through too often to be considered. The
short slot, or B.B.J. pattern is a little better, but if you
are going to mount excluders, it will save you a lot of trouble
if you buy Waldrons. A day in the solar extractor cleans
them up. They are more expensive, but last better than
mounted zinc.

Transport

If you are going to do a lot of dashing about with bees,
a trailer is a necessity. Personally, I have all I can do to
manage a score at weekends and no time left to go gallavan-
ting after blooming crops. Fruit pollination takes place at a
quieter time of year and that I can manage. I always wish
every August that I had time to go to the heather, and when
extracting time arrives, I am always very thankful I have
not got heather honey to contend with too. I hire a lorry and
the lorry driver helps to load and unload the hives. I promise
him a jar of honey for every sting he gets, and, although I
always give him a little honey in addition to his charges, it
has only been " earned " by one driver, who would not run
away when I told him to.

Wire Mesh Screens

These are mentioned in Chapter 10.

Manipulating Cloths

I have only once seen these things used. Although I am
informed that they can be used in a way which saves time,
I use no covering for an ordinary inspection and prefer to
use a crownboard if a stock has to be left temporarily.

CHAPTER IV

THE APIARY

THE most convenient place is the bottom of the garden. I
started like that, but before long I rented a corner of a field
across the road. Most people who have tried it will agree
that about four or five hives is the maximum number that
it is " reasonable " to have in a small garden, and even then
there will be difficulties. There are sure to be times when
even the best tempered bees prefer the garden to themselves ; there are equally sure to be " accidents." The bread
and other commodities may get no further than the front
gate, and, let us face it, bees are usually unpopular (except
after harvest) and will, like the dearest friends, be less so if
absent.

If the apiary is too far from the house to fetch and carry
by wheelbarrow, it is best to put up a hut of a fair size.
This will not cost much more than a trailer and saves hours
of time trundling supers back and forth. Heating and light-
ing can be by Calor gas or oil unless electricity is handy,
and the only trouble is water at extracting time, except of
course, the job of making the place bee tight. It is far, far
better for a person bee-keeping in his spare time to have
his bees within wheelbarrowing distance even at a high rent,
unless the pasturage is very poor. The best alternative is to
rent a piece of land together with an outbuilding on a farm.

Needless to say, the apiary hut on the site is far more of
a necessity for those of us who go about on push or motor
bicycles.

I have read of the "advantages " of having an out-apiary
in the depths of the country in quite a number of books and
magazines. The city dweller may like to keep two or three
hives with a friend in the country (who may not remain a
friend after a season's bee-keeping), and the professional
must have out-apiaries and the organisation to run them,
but for the weekend or small time bee-keeper to keep bees
at a distance from some sort of headquarters is a most unprofitable undertaking. If you live in a town, by all means
keep bees either in the town or in an out-apiary, but neither

alternative is likely to be profitable unless you have plenty
of time to spare for travelling.

As to the site, no end of writers have expressed themselves
about ideal sites and their ideals are seldom alike. You and
I will have to put up with whatever site we can get. Try and
arrange that the hedge between the hives and the road is not
clipped low. Place the hives with their entrances away from
the prevailing wind. If there is a choice of places in a
meadow, it is nice to have the hives on the lee side of the
hedge and nicer still to have them near the gate.

Hive Stands

Two pieces of 2″ × 2″ wood 4 feet long and set 15″ apart
and 6″ off the ground per pair of hives is what is wanted. I
set the hive rails on concrete blocks 12″ × 9″ × 6″, which
are produced in numbers by builders for constructing farm
outbuildings ; they used to cost 1/- each. They had to be
levelled a few times in the first year, and have since kept
fairly level. Better than that is said to be driving in wooden
posts and fixing the rails along the top. I funked the job of
doing this and consoled myself with thoughts of the awful
mess there would be if a post broke. The aim of course, is
to have the stands firm, and the hives clear of the ground
and in pairs.

Fencing

If you are not a farmer, ask a farmer if your fence is
cattle-proof. When you have done it again, you will sleep
better.

Grass

A good dressing of weed-killer when the bees are not flying
or are away at the orchards is a help, but somehow, the grass
that grows in front of hives is tenacious stuff. Sodium
chlorate is about the best weed-killer. An old plank placed
in front of each pair of hives is a help. Corrugated iron
sheets are better.

CHAPTER V

THE MANAGEMENT

Part (1) AUTUMN and WINTER

AUTUMN management begins after the honey is extracted.

I am going to digress for a moment: The supers after extraction can be stored as they are—" wet," or can be returned to the bees to be cleaned up, and then stored "dry." If I had a few hundred hives and two or three times that number of supers had to be carried out, put on, left a few days, cleared again, removed and stored, and all this fitted in between extracting, feeding and dealing with (resulting) robbing, I should take to drink in a big way. Storing combs wet has the advantage of discouraging wax moth without worrying about with noxious smells, and it is said to have a disadvantage in that the "wet" ferments in winter and is not good for bees in spring. I must say that I prefer to store "dry" mainly because, like so many weekend men, I have no beetight storage ; if I stack a pile of wet supers in the garage, too many bees get buzzing round them and nobody dare go in and get out the car. By the way, this word "dry" does not mean that the frames will all be always bone dry or anything like it, but they will have dry woodwork and will hardly "drip" at all.

When extracting is over, I go out into the night and pile up the wet supers on to as few hives as I can ; if I remember to take a box to stand on, ten supers per hive can be managed, but I wish it would not blow a gale the next day almost every year. The supers are stacked up on the strongest stocks with constricted entrances. A week later, escapes are put into the crownboards of the weakest lots and the supers, which have been piled up on the strongest lots are then "cleared" into them, a proceeding which evens out populations and has not yet led to any fighting. If there is any foul brood about, I suppose it would show up in these guinea-pig colonies. These actions have never yet led to robbing although they sound risky, but it is well to keep all entrances very small during feeding and transfers of population.

I can tell from my notes which hives are short of stores and I start feeding those first. Later, a large sheet of squared paper is pinned on a sheet of 3-ply, the hive numbers go down the left and the weekend dates along the top. The hives are weighed by hooking a spring balance under one side, then under the other, the readings totalled and entered on the sheet under the date and against the hive number. Once again, a bit of extra fuss that no doubt the bee-farmer would not contemplate, but which can be done in the evenings by the weekend man and it can be decided in advance which hives are to be given feeders next. I put "f" or "ff" against those hives needing feed normally or extra, and "Fed" when it has been done. I like to get each hive about 90lbs including its shallow roof (there is little difference between a Dadant and a doubled National for this purpose), but, as the lightest hive I have wintered successfully weighed 54lbs in November—I was desperately short of sugar for the winter 1948/49—and the heaviest was over double that weight in the autumn of 1950, it can be seen there is plenty of margin *provided* you use large brood chambers. Keep the roof on while weighing, as it causes less disturbance and unless you have an enormous number of hives with no end of different roofs, it is simpler to "correct" for odd roofs when weighing in the evenings than to put on a veil ; bees come out of a feed hole after the weigher when they will stay in if the roof is on. This method of weighing is not "dead" accurate, i.e. it will not do to record precise gains in daily weight during a honey flow, but it is quite good enough for checking Autumn stores.

The actual mixing of the sugar is a pest. In milk bars there are little electric motors ; a beaker of milk, ice cream, and other ingredients is put under them and a little spinner whirrs the stuff into a foamy drink in no time at all. I have always thought these things would be so useful for mixing syrup. However, I have not got one so this is what I recommend. Tip the required quantity of hot water into a few 28lb tins and add most of the sugar to each tin. Grip the tin between the feet and stir with something like a broomstick, going from tin to tin until all are fairly fluid. Try using a wooden spoon if you like ; it is much harder work. Add the rest of the sugar and continue stirring the tins in turn. If you get fed up with it, leave it a bit and go back to it later. The hotter the water and the weaker the mixture, the quicker the job. " When the sugar is dissolved," which means when the syrup is clear and only a sprinkling of grains show a disinclination to disappear, leave the tins to settle for half an hour before filling the feeders. The undissolved grains stay behind in the bottom of the tins and get used up next time instead of solidifying on the bottom of the Miller feeder, or blocking up the holes of a friction lid feeder. The quickest feeding is done by putting in hot syrup, but to avoid robbing this needs to be done in the late evening. It seems to be agreed that two pounds of sugar to a pint of water is about the strongest solution which will give satisfactory results, but it is all right if you make syrup a little stronger, provided you let it stand for half an hour to allow any granules fall to the bottom of the tin.

Thymol is so little trouble to add that I do not know why it is not always used. I wish it did not have to be dissolved in alcohol, because then the M.O.F. could add the crystals to the sugar supplied on permits and once and for all dispose of the idea that bee-keepers make jam with their bee-sugar, and also dispose of those bee-keepers who do so. The recipe is ½oz crystals in 1 oz alcohol, and put half a dozen drops out of a fountain pen filler or dropper into each 28lb tin of mixture. Do not worry about a few drops more or less, and put it in last thing before feeding. I am informed that thymol may increase the *tendency* to rob. Sugar with *Octosan* effectively stops bee sugar being used for jam and it is supplied cheaply in some countries. I do not know why it is not made compulsory in England and issued without permits.

A good general rule for autumn feeding is to feed at least 5lbs to each stock whatever its weight. After that the balance of sugar can be rationed out to bring up the weight of the lighter stocks. Those that have had the job of cleaning up "wet" supers will not need quite so much.

Small stocks can be strengthened as already stated, by clearing supers into them from strong stocks. A little brood spreading, if you have any empty combs, can be done, but see there is good pollen in the comb used, and feed several times. Generally, this is best avoided, and also any altering of comb order in the autumn, but there are occasions when

it is preferable to take a risk to increase autumn breeding
provided the weather is warm.

Mating Nuclei

When these are finished with, frames and bees can be put
into National bodies and united with each other or with
existing Nationals, after seeing that only one queen is left in
each pile. Mating nuclei can be united with advantage to
"small" stocks which would be the better for the brood in
the nucleus, the few bees and the young queen. Simply remove the old queen, unite with newspaper, leave until the
brood has completed a cycle, and then feed. As the purpose
of autumn management is to go into the winter with heavy
stocks, good queens and plenty of young bees, a few nuclei
can be used to help along the stocks that need help. I might
add that if you unite a number of mating nuclei, wait a
couple of days, re-arrange the frames to give the full com-
plement per brood box and generally monkey about with the
" stock," it sometimes happens that the queen cannot be
found and queen cells can—so be careful. I suggest that the
nuclei be united and a week later, the frames be arranged,
the queen caged and a frame and top feeder filled.

The date of feeding is worthy of mention. " Start " with
the small lots and give them a little if there is time to fiddle
about ; that is, a little given quickly in a big feeder to get
the queen laying. A small lot needs to breed late to get a
population large enough to winter. A big strong stock can
wait until late October most years, provided stores are not
short, a statement which I suppose had better be qualified.
If you have a big broodchamber, you may take risks about
autumn feeding which just are not "on" with little broodchambers. Getting a good sized collection of bees for wintering together with enough stores and their brood on to ten
small frames is a job which can be done and is done, but it
takes a lot of skill and care or else the stock is as weak in
the spring as the majority of " beginners' " stocks in this
country usually are. If you have a big brood body with a
strong stock and a moderate larder, nearly always, it will
winter without attention, and the attention it gets in the
autumn contributes to the value of the stock in the spring.

So, if the stock needs a few more bees, feed early and in instalments ; if it is strong, feed late. If stores are short in the hive and the sugar ration has not been boosted, feed late. Late feeding usually results in unsealed stores which is only troublesome if thymol is omitted.

Candy

Too much has been written about this stuff already. It seems agreed by those who make bee-keeping their business, that it is not worth while. If you have little broodchambers and not enough stores in them, it may be necessary to use candy if the M.O.F. produces a bonus ration really too late to feed properly. Apart from that, feeding candy means a lot of work and it is debatable whether any good whatever, results from its use, ever. It· has been agreed that candy is no " stronger " than honey, so no water is needed to dilute it. Maybe it is not " stronger," but it makes life difficult for the bees to pick-axe it out; and honey attracts its own water and candy does not. Hydroscopic is the word.

Do not try to invert syrup or candy with medicaments. There is some difference between chemical and animal inversion and the bees will only have to bring it back to normal and invert it their own way. The bio-chemists talk about laevulose L and laevulose R, so maybe the simple way to explain the matter is to say that if you want to go from L to R you have to go the whole way round through zero again, so it is easier to let the bees start at zero in the first place.

When feeding is complete and the feeders put away, the bee-keeper's wife may 'heave a sigh of relief and make a list of all the odd jobs about the house which have been neglected for months. In fact, there is little else to do. Remove entrance blocks which have the little entrance in use, and put them round the other way. In doing so, pin a strip of 3 ply along in·order to reduce the height of the entrance so the bees can just march in and the mice cannot. A strip of perforated zinc on the entrance block does as well, and so does a whole strip of zinc instead of an entrance block, with a low doorway in the middle. Arrange for some form of top ventilation. Complete the last weighing in early November, and put it in the hive record book. If you weigh again in early February, it may surprise you to find an average of a mere 5 or 6 lbs loss of weight, thereby proving that the bees would have to be desperately short of stores before feeling the pinch. The loss of weight in February alone varies between 6 and 20lbs., and if the weather is warm enough for a high rate of consumption, it is warm enough to feed the starving ones.

It has often been said that if you have not got enough sugar, you should "leave" the bees some honey or "return" some honey. My earnest advice is—don't. If you have large broodchambers, this will not arise, and you can always make sure it does not by supering about one hive in ten with brood frames so you can supplement light hives by adding a frame of honey. If you leave a super on for the winter, it means that those frames cease to perform the functions of extracting frames, and become brood frames. They can be put back to their original function next year, of course, but will always be harder to uncap and will tend to be used for pollen more than others. Also, these frames were not made for brood rearing and are normally incorrectly spaced for that purpose. If you must do it, remember to remove the excluder. Better is to ask another to "winter a couple of stocks for you" which is a legal way of borrowing a little bee sugar from a neighbour with some to spare.

Put a brick on shallow roofs and leave alone until after Christmas. If you have afforded deep roofs, you do not need to use a brick.

The next job is to see that all stands are still level. Move hives off uneven stands and get the lot level while the bees are quiescent. Any rearrangement of hives on existing stands or on to fresh ones is best done in January.

Winter losses get a page or two in most books, but, personally, I know little about it. I lost none last winter, one the winter before, two the winter before, and none in the other years. This is about 3% and I think it is quite a good average for the East Coast. I am sure a large broodchamber makes for better wintering. Also, it is important to leave the hole in the crownboard open and not stuffed up with sacking, and put blocks of wood on two corners of the crownboard to leave a good space under the roof.

If you have supered with brood frames in order to get them drawn, and then after extracting, decide to double the hive

in the autumn, do not forget that bees like their stores above
the cluster. So put the empty frames underneath, or you
are likely to lose the stock. There is no point in doubling in
the Autumn unless the stock is sufficiently strong in stores
and bees to occupy the added broodchamber ; if not, better
store the spare broodchamber with the supers.

I do not like to leave this chapter without a mention of
wasps. If you want to destroy a nest, an old farmer gave me
a tip. Dissolve cyanide in water and keep stoppered. When
a nest is found, spear a bobble of cotton wool on a pair of
scissor points, wet it in the solution and pop it in the hole.
Efficacious in every case, and a little cyanide does for no end
of nests. Another tip, let it be known among the small children in the neighbourhood that a wasp's nest reported within
200 yards of the apiary is rewarded with a (½lb.) pot of
honey. This tip does not always work—small children like
honey, but the local ones would rather look for birds nests
to tip up, or so it seems to me. There is a Plant Protection
product called "Waspend" which I hear is simple to use and
really does the trick. A doctor said to me " if people only
knew just how dangerous cyanide is, they would not allow
any in the house."

Part (2) SPRING and SUMMER

The Management

I WAS once asked a question at a lecture, "How can you
really learn anything from books when each author tells you
to do something different?" I was ass enough to reply with
a string of names of "reliable" authors and the meeting
was at once in uproar. It is by no means an easy question
to answer, but I will try to do so now. I was lamenting
some of my troubles to a commercial bee-keeper and his
reply was to the effect that my troubles and problems were
very like his. We both had full time jobs, mine in an office,
his in various out-apiaries and that my apiary was just like
one of his smaller out-apiaries in that it could only be visited
at set times in the schedule and not just whenever one felt
like it. In other words, I spend half a day a week (trying

to) work an apiary in the same way that a commercial man may allocate half a day a week or per nine days to one particular out-apiary. The answer, therefore, to the question I was asked is that *if* you are trying to work a small apiary as a commercial man would, or rather " in a proper business-like manner," *then* read all books for their educational value, but follow the recommendations of those authors who have *made a success of honey production* on at least a moderate scale. Honey production, please note, as opposed to income from bee-keeping, which can be an elastic term and might include such things as royalties on books, lecture fees or even gifts from grateful bee-keepers. If, on the other hand, you are managing a small apiary, maybe of W.B.C. hives, have plenty of time to fiddle with them, and are keeping them for honey production mainly, but to quite a large extent for the study of insect life, botany, etc., to provide fuel for a microscope, or just, to group a variety of reasons together, because you are plain " interested in bees," then the answer must be rather more vague. Still read all the books, and the reviews in the bee press to check up on any book that may give quite bad advice, and when you have got along with bees for a year or two, you will find you will soon get an idea of the system of management which will suit your scheme and in which you would be interested to experiment. There is no need to get confused because two authors recommend two entirely different things ; both things will probably be quite all right, but one will suit one man and not the other. More important, is to work out for yourself why both things work (or should work), and when you have done that you will be more knowledgeable, and that is a thing which counts. I am, however, writing for those who would try to keep bees in a fairly businesslike manner.

In East Anglia, spring for the bee-keeper may be said to begin with coltsfoot and dandelion, continues with willow, and with the apple blossom it merges into summer.

It is really a debatable subject whether there is any such thing as spring management. Many books say you've got to feed, and having got that off the chest, follow with a dissertation on checking over equipment. In practice, there is very little to do indeed. Once the weather begins to warm

up, all stocks should be bringing in pollen and it is useful
to make a note of those which are not, especially if they
are not generating any heat. Even that is only " useful "
to the extent that when the first examination takes place
there is a note that trouble is likely to be in certain particular hives which are better dealt with early on.

Floorboards should be changed about February. Remove
the dirty floorboard and substitute a clean one. Have a look
at the old floor in case there are indications that mice are
there or the bees are dead. If the colony has died out, close
up and remove the hive and try to discover why it has
died. Send in a comb for examination if brood disease is
even suspected, and also send in some bees if any living.
In the case of mice, get them (all) out first, remove the
nest and damaged frames •if the cluster is well clear and
give a comb of stores if needed.

While I am on the subject of feeding, I may as well finish
it. Spring feeding is necessary if the hives are light at any
time in the spring from whatever cause. If the " cause "
is high speed breeding unaccompanied by incoming nectar,
feed a strong mixture. Apart from· this, a weak mixture is
good enough. For stimulation to increase the strength of
the colony, it has been fairly conclusively proved that water
is more important than sugar. Presumably the stocks
dwindle in spring because the mortality rate among water
carriers is high, so if water can be provided in the hive, the
water carriers can perform the useful job of keeping the
brood warm instead of dashing out and dying of cold or
getting drowned. Do *not* add salt to the water. Whether
bees like salty water better or whether they do not, it may
be good for us, but it is very bad for them. One man's
meat, etc. My bees do not go up into my overall feeders and
fetch down water, so I give them a little sugar with it. I
hear they take it better out of honey tins with holes knocked
· in the lids.   ·

I usually go and have a look at the hives when early
spring flights begin. Quite often a number of bees will be
found clinging to the tops of grasses unable to get back and
with their wings stuck out at queer angles. I collect samples
of these in matchboxes and have a look for acarine. I never
have found acarine in my bees yet, but these early fliers

usually show the K-wing sympton, so I usually send some away to have the diagnosis checked. I cannot say for certain that it is a good time to test or whether it is a good way to take a sample, but to me it makes sense. If acarine is in one or two hives, it would be nice to know in February.

In this part of England, there is nothing to be gained by having what has been called " the first peep within " until the bees are working the Willow *(Salix fragalis)* and the day is warm. The Sallow, Goat Willow or Pussy Willow *(Salix Caprea)* produces its catkins earlier and should not be confused with Crack Willow or White Willow. If it is an incredibly early spring like 1948, it may be advantageous to have an earlier inspection in case a number of wet and cold weekends follow ; it may éven be that supering is required if you keep tiny hives. The purpose of very early inspections seems to be to content the impatient and that is all. Once the willow is being worked, it is time to think about supering those hives (if any) which are going to the orchards, but the rest of the hives could perfectly well wait until the end of April in most years.

Most of my hives go to apple orchards for the blossom time, so I regard summer as beginning when I get a telephone call to say " I'm spraying with lead-arsenate the day after tomorrow," and then I know I have to get my hives back. I have lost swarms at the fruit and had to collect others, and quite a number have been collected for me, but I regard this evil as slightly less than having to examine hives which have their supers stapled on and then hammer the staples in again afterwards. Besides this, I cannot get round a score of stapled-up hives in a couple of orchards in half a day and nor can anyone else I know. Going to the fruit is a perfect pest in every possible way, and if it is not made financially well worthwhile, it is far better left alone. The fetching and carrying are bad enough, stapling takes an hour in theory and two in practice, some hives come back weak and some dead, and more often than not, a number have ripe queen cells. I am thankful when they are safely back and summer has begun.

There are already almost too many books devoting half their contents to summer management and especially swarm

control. I am going to assume the reader has a fair knowledge of swarm control, and of the various controversial
issues, side-issues, fashions, and factions, and confine my
remarks largely to those points in management which I
think are suitable to weekend bee-keepers.

The objective of good management for honey production
can be summed up as :—

    1. There should be as many bees of foraging age as
       possible at the start of the honey flow.

    2. The foraging bees should forage.

    3. The incoming nectar should not be used to rear
       grubs which will not be bees in time to be useful.

It sounds all very obvious. The difficulty is to get the
maximum number of foraging bees at the right time without having the stock, as Wadey expresses it, " running to
seed by swarming." One point I wish to make is that it *is*
good bee-keeping to have a good total crop compared with
other producers, and it *is not* necessarily bad bee-keeping
if this total crop is produced by only half the hives.

If there is a late summer flow, hives that are not strong
enough to make " good " use of it can be reduced to nucleus
strength and have their surplus bees added to a stronger
stock which *will* put them to good use. Just move the
small lots aside to lose their foragers, to the adjacent hive,
and take their supers, bees and all, and add to a strong
lot. Then check that there is not too much brood left for
them to cover. In the case of a summer flow, more fore-
sight is needed as a stock which is coming along nicely
does not want to be put back so it cannot take advantage
of any other flow later in the year. But it must be remembered that removing bees from a stock in appreciable
numbers reduces the queen's rate of lay, so this is advantageous *only when* those eggs would be laid at a time of
year when brood rearing is better restricted.

Uniting : Manley says there is very little of this on a bee
farm if you know your job. Note : on a bee farm. Two
stocks can be united during a honey flow, and that they
will give more surplus combined than both would sepa-
rately is well known. Do not worry about halving the
number of stocks by uniting. After the flow, the powerful
lot can spare plenty of sealed brood which will build up a

nucleus used for queen mating quickly enough for it to be in a good state for wintering.

Snelgrove.

It seems to have become fashionable to cry down the Snelgrove swarm control system. In one book I find that it is no use to the bee farmer, which is an unfortunate remark as it is almost certainly not true. I agree that it is unsuitable for a bee farm with Dadant hives, and I also agree that it would probably not be best to Snelgrove all hives. Snelgroving uses up more time in late May and less in late June, so it appears to depend upon how each bee-keeper is fixed for time in these months, and it also makes a difference if there is good weather for manipulations in May. Whether that apiary will give as much average yield as the other is arguable, but it is also very debatable whether it is a paying proposition if National hives are the sort used. I also read that ample room in the apiary is required as bees fly out on all sides, an objection I do not follow as I find a six inch space between hives gives enough room for a bee to fly about in. I do not use it " by the book " partly because it is not so good with Dadants, and because it is better with drone comb supers which do not get so cluttered up with pollen.

The most serious objection is summed up best by Wadey who reasons that it is bad for colony morale to split up a hive into two separated parts. Of course, if the hive is going to separate itself by swarming, little harm is done, but the modification whereby the system is used to " cure " swarming as opposed to " discourage " it, does not seem to have caught on, and one does not know in May which hives are going to swarm in June.

I have said I do not follow the book, but I do use the " board " sometimes. It is often that one hears of " failures " with the method, and the answer is always " you did not follow the book." As far as I know, few people do " follow the book," a proceeding which is not as easy as it sounds because the stock is supposed to be in a particular condition before the operation. However, here are some other uses for the board.

If a stock has been Snelgroved, a week later the upper storey will be moderately strong as bees have been emerging in it for a week. Another brood box, with preferably a preponderance of sealed brood, can be added to it, but do *not* work the wedges for another week or the extra brood may be chilled. The removal of a lot of brood from a hive is sometimes desirable and an extra box above a Snelgrove board is not a bad place to dump it in. Work the wedges the day before the virgin emerges.

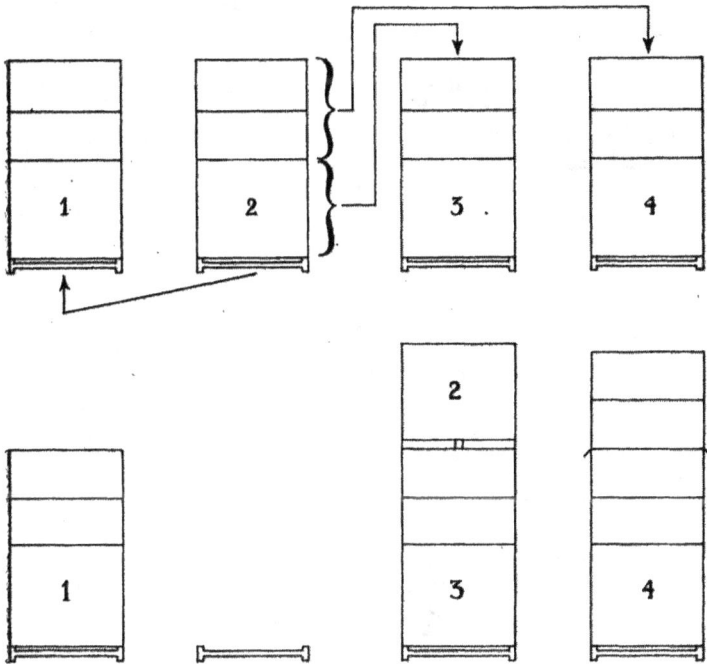

Method of dealing with a hive (No. 2) preparing to swarm

Hive No. 1 gets the flying bees.

Hive No. 3 has brood box of No. 2 placed above supers on Snelgrove board.

Hive No. 4 gets the supers from No. 2 together with the bees in them. Unite with newspaper.

If one of two stocks on the same stand is preparing to swarm, a good way of dealing with it is to put a Snelgrove board on top of another stock, and just put the swarmy lot on top of it until it ceases to be " broody." The super can go on a third hive, bees and all, if you like, and the flying bees all go into the adjacent stock. It has been said that this switching of flying bees with minds in a mood for swarming into a normal stock will " infect " that stock with the other's disease of the mind. This may be so. I will not be categorical about it and will just say first that I have not found it so, and secondly, that I do not see how they could. I think it was Wadey who first described the phenomenon of hive organization as the " hive mind," and as far as I can see once bees leave one hive they get absorbed into the hive mind of the next colony and *vice versa* and they certainly have not got enough grey matter to spread rumours of free outings around the feminine members of another stock. As for the treatment of the colony which has been put on top of a Snelgrove board, much depends on circumstances. The sealed cells can be destroyed and the bees usually destroy the rest, but if not, it is as well to dequeen and introduce a ripe cell. Work the wedges the day before the virgin emerges. If it fits in with the scheme of things, it is a splendid place to have a stock that wants requeening as the foragers can be deflected in a moment. When the new queen is mated and laying, there is one excellent use to which her stock can be put. It will be referred to as " U." If two colonies on the same stand start preparations for swarming, remove their supers and unite to " U " with newspapers or merely by leaving the supers about for the air and light to get at them for 5-10 minutes, it depends on the weather, robbing, etc. Then remove the two swarming lots and put " U " in the middle of that stand. Cage the young queen in " U." You can add a lot of sealed brood from the swarmy lots at the same time, but be sure not to add any queen cells. If there is a flow on, it is wonderful how much honey there will be in that lot the following week. At this stage, I must make it clear that these and other operations are not put forward as " methods to use in the circumstances indicated," but rather as " methods which might be considered and accepted or rejected according to the conditions in the other hives, the time of year, the time when the local honey producing plants are at their best, of course the weather, and all the rest of the factors."

Another handy " use " for the Snelgrove Board is this. Leave as much brood as possible in the lower box with the queen, and put the rest of the frames (which should contain some young brood and eggs for a queen, and sealed brood for a virgin) in the upper box. Shake plenty of bees into the top box. Put on the board at once and cage a queen if you have one and a virgin if not. The position now is that the top box is a " nucleus " and the bottom box is rather crowded with brood. Leave the hive alone for a week (or longer for a virgin) and then reverse the boxes. This should put an end to the year's swarming for that stock. A further point is that a week later, the top box can be taken away and if the Snelgrove entrance was in front, the flyers will soon find the proper entrance. Almost simpler is to remove the queen and all combs, just leaving the upper lot of bees in an empty queenless box from which they usually trek off down the hive front and join up with the lower lot after some antennae-waggling. If they do not march down, shake them in front of the main stock (or any other stock) *after dark*.

By now, you will have realised that I have no " system " of swarm control, but plenty of methods of management. The basis of these methods is the three points I made earlier and which can now be examined in more detail.

1. There should be as many bees of foraging age as possible at the start of the honey flow.

We all wish we knew when and if it was going to "flow." However, if it flows early, it is just a question of autumn management and wintering good strong colonies with good queens. Apart from the early flows, if any, the only way to get foraging bees is to have as many queens as possible lay as many eggs as possible up to the end of June. Now please do not fall into the too common fallacy of saying immediately "an egg laid on 30th June will not produce a forager before mid-August by which time all will be over bar the shouting. ' That is nonsense. The bee in question will emerge on the 20th July and will take over internal duties, reception of nectar, etc., and so " relieve " another bee which can take on field work at an age younger than that at which it otherwise would.

It is no good aiming for a hive full of foragers ; you need a good big supporting team too.

As many queens as possible is difficult to achieve without stock division which reduces individual output. Of course, it is necessary to do a little stock division to start queen rearing, but that is a necessary evil and it is no help to the harvest. We can, however, do something to ensure that our queens really lay hard.

A good quality queen will usually lay at less than " normal top speed " if the bees are preparing to swarm,' if there is continued dearth of nectar, or if the stock is weak. She will lay at more than " normal top speed " if the stock is competing with another (see *The Skyscraper Hive*) and if it is stronger or " warmer " than normal early in the year. We can do something about *increasing* the rate of laying by superimposing colonies in the spring with or without ventilated boards, but whether this is more nuisance than it is worth is a matter for each bee-keeper to decide for himself. Weak stocks—perhaps I should say small stocks—can be placed on top of others where the extra heat to some extent compensates for the lack of bees. As already said, I find Snelgrove boards useful barriers for this manipulation. Bees preparing to swarm reduce the laying rate of their queens and that point will be dealt with later.

2. The foraging bees should forage.

They won't, the experts say, if they are trying to swarm and the bee-keeper is thwarting them. They won't if there is a shortage of house bees as some will have to stay at home. They won't if they have acarine. Of course, they won't if there is no nectar, but that is not what I mean.

3. The incoming nectar should not be used to rear grubs which will not be bees in time to be useful.

This, of course, is the old saying " one should raise bees for but not on the honey flow." There are all sorts of systems to counteract this evil, such as de-queening at the start of the flow. This, by the way, is not a bad plan if you introduce (successfully) a virgin at the same time and can replace her later should she take overlong to mate, but it is really risky. Much better, is to avoid as far as possible having any stocks in big broodchambers which are strong in bees, a little short of brood, and containing queens just coming up' to the peak

of laying. Such stocks would be fine for the heather no doubt,
or, to be more precise, fine for any other flow but the one in
progress.

The proper way is to have good normal stocks in which the
queen reaches her laying peak before the main flow.

* * *

It now remains to describe a scheme of management.

The commercial bee-keeper tries to inspect his colonies at
intervals of nine days. Although the weekend bee-keeper
aims at a weekly inspection, the need for clipping queens is
still there unless a member of the family is at hand to catch
all swarms. Even in this case, nobody I have heard of has
seriously attempted to explain why queens should *not* be
clipped. It has been put forward that the disadvantage is
that it does not prevent swarming, but that is an old bit of
nonsense which, tra-la, has nothing to do with the case. I am
not concerned with the artistic arguments against clipping.

The drill is, therefore, to try and get round all the broodchambers each week. If this cannot be done owing to the
weather or other engagements, get round to those in which
trouble is expected and to those which were not examined
the week before. How much " examining " needs to be done
depends upon the condition of each stock, the experience of
the bee-keeper, and, a point I have never seen made
before, the adequacy of the hive record sheet. An experienced
bee-keeper may remove a couple of combs and " know "
perfectly well that he has learned as much from them as he
would if he removed, peered at, and replaced the lot. For
example, if in May you start at one side and remove the
outside comb and find eggs on it, *normally* that stock can
be closed up and a note made that it is OK. The year 1951
was exceptional in this district, and I remember a stock with
eggs in both outside combs and nearer the middle were several
queen cells almost ready for sealing. I thought this was
another example proving that " bees do nothing invariably "
but I found several other stocks in much the same state. If
you find sealed brood first, it merely means that the queen is
on another cycle and is no longer expanding the brood nest
in the direction of the frame removed ; it does not mean that
the bees are preparing to swarm, but, depending on the time
of year, etc., it does mean that the bee-keeper needs to look

further. There are signs which we all come to recognise better with each year of experience, most of these being a sort of " feeling " that preparations for swarming are being made, and as I cannot account for this " feeling " I conclude it must be something in the behaviour of the bees. Each weekend the first few stocks are extensively examined, and after that the examinations become less and less thorough unless there is a spate of swarming.

As the month of May nears its end, we are sure to find some stocks preparing to swarm. " Something " needs to be done about these and there are quite a number of choices. All cells can be destroyed when first discovered, provided none contain advanced larvae, and it may be that a few of such stocks sometimes forget about swarming. More often than not cells are there again the next weekend. The golden rule is then that if the stock is to remain where it is, the queen *must* be removed. One plan is to make up a nucleus with her beside the stock, destroy sealed cells in the stock and close down. A week later, transpose the stock and the nucleus (you may lose a tiny cast that way). But remember any sort of stock division means a reduction in the honey making capacity of the stock.

Another method is to remove the queen and all cells and put in a ripe cell. Another way is to remove all cells and introduce a young mated queen—if you have one and if you can. A *splendid* thing to do is to split the stock up into nuclei for queen mating or use it for cell building. Splitting a stock into nuclei does not always check a swarming tendency however, and there may be trouble with little swarms from the nuclei. In fact, more has been written about this one subject of what to do with a stock that wants to swarm than about any other aspect of bee-keeping.

When hives are kept in pairs, the stock which is preparing to swarm can be picked up and taken to another site in the apiary. Once this has been done we get left with (a) a hive without fliers and (b) fliers without a hive. Regarding (b), either a nucleus with a newly mated queen (cage her) can be put to catch these fliers, or another small stock can be used, or the fliers can be left to find their way into the adjacent hive. Regarding (a), the stock is now denuded of fliers and is comparatively (!) easy to requeen. If the cells are destroyed—and the bees may do this themselves so leave the unsealed ones a week—the stock may be cured. A nucleus can be removed from it, given a " good " cell and left to mate its queen. It may be necessary to remove a frame of brood and give it to another hive as the depleted stock may not be able to cover it all.

I have been told I do rather more " switching " bees than most people, and I am frequently told that this " switching " is risky for two reasons—first because it frequently starts fighting and secondly because " switching " bees from a stock preparing to swarm into a stock that is not, often leaves two stocks preparing to swarm when there was only one before. I cannot say I have noticed either of these two calamities in my apiary, but that does not mean it is not risky with other strains of bees. I have a friend who, I think much too frequently, used to unite stocks. She has lost so many colonies by trying to unite them that she tells me she no longer does it. I have helped to do some of this uniting and she must have tried every known method, but practically every time the bees fight nearly to extinction. That must surely be something to do with her strain of bees, and with her bees I could not expect to get away with some of the possibly risky operations I perform in my apiary without a thought of the consequences.

There is a very real risk of finding a stock queenless after switching bees into it. If a very small stock or a nucleus is transposed with a very strong stock, the queen in the small stock is in danger, especially if she is young. I am of the opinion that it helps if emerging brood is added to a small stock a week before any old bees are added, and that caging the queen in an automatic release cage is only a little help. It is difficult to see why very young queens should be treated worse than old ones by bees from another stock, and until the cause is diagnosed the best treatment must remain a matter of guesswork, trial and error. A principle widely accepted regarding queens is that they are comparatively safe when surrounded by their own attendants, so it seems logical that the thicker the " screen " of attendants the safer the queens should be, and adding ripe brood a week before does thicken the screen.

My own method of management is this. As early as prac-
ticable, I make up a few mating nuclei using a hive preparing to swarm if there is one. As soon as these nuclei have
mated queens, they are put into hives and given some
emerging brood to strengthen them, the brood coming from
stocks likely to swarm. The following week these small stocks
are ready to change places with stocks actually preparing to
swarm. Once I have a stock with a new queen on a stand
receiving a full quota of fliers, the first objective is attained.
The next week that stock is ready to receive more sealed
brood from any stocks that need to lose a bit, and also its
mate on the same stand can be moved away, leaving a very
powerful stock which should be looked at the week after to
see its queen is all right, and then left alone. The moved stock
will probably not swarm and will " come on " for a later
flow.

This method is, of course, a race against time. It is one
long struggle to get young queens into strong stocks before
those stocks think about swarming. Those stocks that prepare to swarm before the young queens are ready, are moved
aside and then "dealt with " one way or another, while
their fliers are put to good use in a honey producing stock.
As a method, I commend it to weekend bee-keepers who can
work out all manner of variants. It needs a certain amount
of work preparing the weekend's work, and rather extensive
note-taking and recording, so it will probably not be much
good for the large scale man.

I have read most of the modern literature on summer
management, and it seems to me that the normally recommended system is to regard each hive as a separate entity.
Each colony is examined from time to time, and various
manipulations are carried out with a view to forestalling
swarming. Should the bees prepare to swarm in spite of the
discouragement, other manipulations of a more serious
character are put in hand. After such management, the apiary
in July has the appearance of a number of tall hives whose
colonies did not attempt to swarm, a few hives of medium
height either because they contain below average quality
queens or because swarming was prevented with only
moderate manipulation, and a few hives with no supers and

satellite nuclei. The latter group are the failures — those stocks which determined to swarm and had to be prevented by means so drastic that the honey crop was reduced, and those which failed to go ahead in the spring.

In the early part of the season, one may remove brood (preferably emerging) from stronger stocks and add it to those which need a boost to bring them up to average. The removal of ripe brood from a strong colony may have the effect of preventing it from swarming at the time when it is most needed for honey production. This manipulation does have the effect of " evening out " the colonies, and if that is considered to be good bee-keeping it is probably the reason for the commonly held theory that an apiary of hives all of a height is a hall mark of efficient management.

There is much in this last paragraph that I cannot agree with.

CHAPTER VI

THE BEES and the QUEENS

It is undoubtedly important to maintain a good strain of bees. Unfortunately, despite their good resolutions, many small bee-keepers fail in this, and it must be admitted that it is not an easy task. Some bee-keepers do work to a plan, but if this is poor, as it so often is, there is no improvement in the strain as the years go by. There are three main reasons for this. First, the difficulty of doing the job compared with other forms of livestock (e.g., the poultry farmer buys a new cock, the cattle farmer rings up the Artificial Insemination Centre). Secondly, the diversity of objects. Thirdly, the uncertainty with regard to drones. The quality of cattle has been improved as a result of legislation ordering the licensing of bulls, but there cannot be any law ordering the suppression of the abominable drones which predominate in this country.

The basic plan should be to breed from a queen of really good strain. This may sound obvious, but there are many more complications than one would expect.

Some books recommend buying a queen " from a reliable breeder " and making a start with her. Unfortunately this advice is difficult to follow. A breeder queen of a really good strain (" Brood Queen " is probably the correct term but might lead to confusion with " Queen Brood," etc.) can only be obtained from a person who has sufficient hives managed for honey production to be able to see the results of his breeding in tons of honey as opposed to numbers of queens reared. The big commercial honey producers are people whose livelihood depends directly upon the quality of the strain of bees kept, and they mostly keep and maintain very fine ones. The commercial queen breeder in this country on the other hand does not depend on the honey-getting propensities of his bees to such an extent, because he also requires to keep a prolific, gentle and beautiful strain. Admittedly his business will suffer if his strain is not also efficient, but while the commercial queen breeder has not only got to keep his eye on the other factors, but also unless he himself keeps a very large number of hives for honey production, he has insufficient means of testing the results of his breeding, and in some cases he takes no theoretically sound steps in the matter at all.

It is not always easy to obtain a first-class queen from a big producer. These men are so busy managing large numbers of hives for honey that they have not time to rear queens except for their own hives, and some of them say they have not time to pack them for posting. The answer appears to be as Mr. Manley says '' You may have to pay a good bit to get her.''

There is no doubt that yellow queens are easier to find, at least until the bee-keeper has had the experience of searching through a few thousand hives for them. There are excellent strains of yellow bees which differ between themselves sufficiently to cater for the needs of most bee-keepers ; it is the strain that counts, not the original habitat of the bee generations ago. There are some people who require a strain suitable for use in a single 10-frame British Standard brood chamber, and others want one for use in a system employing a brood chamber consisting of two Langstroth deep bodies. Some bee-keepers would rather have a strain capable of producing 5 lbs more honey than other strains, even if the strain is bad tempered. To a large extent different strains are required in different types of locality, e.g., if there is no early flow, a strain which builds up early is a nuisance and expensive in sugar. The fruit-grower wants a prolific strain which *will* build up early and breed great quantities of bees in time for the apple blossom, and if these bees tend to excessive swarming later, it is a secondary consideration. The local severity of the winter is another factor. These different qualities can be found in strains of bees of all colours. The dream of '' a '' National Queen Breeding Centre may be excellent and certainly better than nothing, but the Centre would have to produce queens suitable for the needs of different bee-keepers who demand different characteristics and colours. Until instrumental insemination is practical on a commercial scale in this country, each strain and each colour needs to be bred in a place apart. It is a waste of time to preach that one strain par excellence is *the* strain for all bee-keepers. It is the equivalent of suggesting to farmers that one good breed of cattle is all that is required. Even if it were so, such a statement would get a poor reception ! So, assuming that honey-getting is the main quality desired, it is first necessary to find an efficient honey producer who uses the same sized brood chamber as yourself, and has bees with the qualities desired, and then buy a breeder queen from him by "paying a good bit to get her." There may be many factors to take into consideration such as gentleness, but the one that *really* counts is whether the strain shows a good net profit for the owner *from the sale of honey.*

In many books the amateur queen breeder is told to rear queens "from his best stock." This sounds good advice, but may easily be thoroughly bad practice, and is definitely bad theory both for reasons of environment and genetics (heredity).

Consider environment first. How often have we all heard a bee-keeper say "That stock in the old hive in the corner always gives the most honey—I must re-queen the other hives from that lot sometime." It is well known that a wind-break will shelter hives in certain positions in relation to it, but it may also have a very bad effect on other hives in nearby positions. I read of a bee-keeper who erected a wind-break and found that the hives in the third row seldom wintered well ; presumably they got an extra blast from the wind curling down over the top of the windbreak. With natural windbreaks one hive may be placed in a far more favourable position than others quite close by. Apart from this the best honey-producing stock in an amateur's apiary may well·be the most docile, because the vicious stocks get comparatively little attention. Alternatively it may be the vicious stock which gives most because it is mauled about less !

Now consider heredity. The procedure of breeding from the best stock is often unsound. The reason is that the first cross of two pure types is probably good for honey produc-tion. To take an example, a motley collection of mares of all colours and sizes mated to an equally motley collection of donkeys, will produce mules which will (with very few exceptions) look as though they had been cast in the same mould. Similarly with hens : the most usual sex-linked cross, a Rhode Island Red Cock × Light Sussex Hen will produce a splendid collection of yellow hens all looking alike, all fine birds, and all good layers. This is known as the F.1 generation, and woe betide the poultryman who breeds from these fine birds because their progeny will be all shapes and sizes and most erratic layers. With bees a first cross (F.1) between two pure breeds should therefore be very good for honey, but even if it is the best stock in the apiary it would be the worst to breed from. It may be argued that what applies to nearly all birds and beasts does not necessarily hold good for insects, and although a lot more work requires to be done on the subject, the indications are fairly clear, and I am told by big honey producers that breeding from the best honey-getting stocks leads to failure more often than not.

SUMMARY.	Offspring : points for—	
	Efficiency	As Breeders
Breed Pure to Pure (same race)	8 ...	9
,, ,, ,, ,, (diff. ,,)	9 ...	1
,, ,, ,, Hybrid ...	8 ...	4
,, Hybrid to Hybrid ...	2 ...	1

So if we take it for granted that the best honey producing stock in an average apiary is headed by a queen mated to a drone of a different type, it follows that this excellent stock with its huge pile of supers is almost certainly a bad stock to breed from. To take another example from chickens : at the Experimental Station in Ohio all the birds were trap-nested so the laying records of each hen were accurately known. Each year only the eggs from those hens with the very highest egg records were hatched. After some years of this it was clear that the experiment was failing, and the egg production was no better at the end than it was at the beginning. People were dumbfounded, but the answer was really quite simple : the hens with the best egg production records did not necessarily have the best egg production daughters. The poultry breeders now know that the way to breed is to compare the records of the *offspring* of selected cocks. The flock (not hen) with the highest record was (they argue) sired by the best cock, and it is from that flock that they select cocks for the following year's tests. This system works !

Geneticists say that the first people to consult them were plant breeders, and great strides have been made in breeding strains of clovers, wheat, peas, etc. The poultry breeders are beginning to take their advice. Many racehorse breeders have worked on sounder lines than most stock breeders for years : but it is not the Derby Winner which should command the astronomical stud fee, but the stallion that sired the winner. Derby winners do however, also command enormous stud fees, although many of them are failures as sires. Now the geneticists are trying to tell the cattle breeders that a bull is valueless until the records of his get are known to be good. The " proven bull " has to be some six years old before his daughters' records are known, and by then he has usually developed a foul temper which may account for the unpopularity of the scheme.

It is the same with bees, but as usual rather more difficult. It does not matter how much honey the breeder queen's stock stores (except that a poor queen is " unlikely " to have good daughters). What matters is how much honey is stored by her *daughter* queen's stock. The fundamental difficulty in determining this is that whereas the cow produces milk which can be measured and the chicken eggs which can be counted, the queen bee produces no honey and neither do her brothers. So the daughter queens mate with different drones, and the workers which get the honey are only half from the breeder queen. If you breed half a dozen virgins from a selected stock, it takes precious few mismatings to turn the most careful records into useless rubbish. To do any good you need comparative results on a very large scale. Manley says that he does not like using a queen as a breeder without trying out her virgins first, but to this I think must be added " . . . . without trying out a good number of her virgins first," and this method of selection must be continued each year. So it can be seen that because the value of a queen cannot be measured directly (since she does not produce honey) it is even more necessary to work on a large scale with bees than with other creatures. There are many special difficulties peculiar to breeding bees against which we can set one advantage—we do not get asked a fancy price for a queen because she is descended from one which won first prize for good looks at a National Show.

It has been said that colour in bees is determined by
single " colour genes," as it is in budgerigars and in some
breeds of chickens, dogs, etc. I wish this was the case.
Geneticists know that when colour is the result of one gene
it is possible to breed that colour into a different breed in a
few generations, e.g. it would be possible to breed a golden
Caucasian, a black Cyprian or a grey Italian. The method of
adding one gene to a breed can be simply explained, which
will be done later. If you take a creature with a white colour
(given it by one gene) and one of a black colour (also one
gene), the children will be grey creatures unless one colour
is dominant. Mate the greys amongst themselves, and the
result averages out at one black and one white for every 2
greys, or " khakis " if you follow the rhyme about the
fellow called Starky which is too well known to need re-
peating. His peculiar children showed the colours in the first
generation when it should have been in the second, and in
any case is nonsense because human skin colour (unlike
human eye colour) is not governed by one gene. If either
the black or the white colour is " dominant " the first (F.1)
generation will be of the dominant colour, and of the second
generation, three quarters will be dominant and one quarter
recessive.

Supposing one gene gives a black colour, and if the gene
is absent, the creature is white, the result can be shown in
diagram form. The " black gene " is shown as " B " and
its absence as " b." Genes go in pairs and each child takes
one gene from each parent.

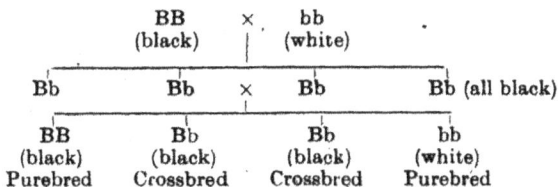

BB × bb
(black) (white)

Bb Bb × Bb Bb (all black)

BB Bb Bb bb
(black) (black) (black) (white)
Purebred Crossbred Crossbred Purebred

As drones are " haploid " creatures, their genes are
single and not in pairs, a point which complicates the diagram : as usual, bees are difficult ! The queens however have
double (paired) genes, and " supposing " yellow colour is
given by one gene, the result of mating a yellow queen (YY)

to the black drone shown as " y " (i.e. not having the gene which gives the yellow colour) would be as follows. Note : queens and workers take one gene from the parent drone and one of the mother queen's genes only.

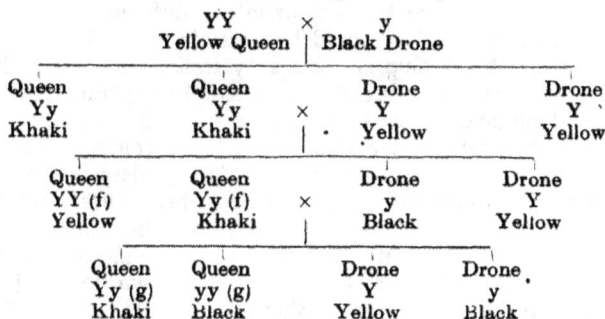

$$\begin{array}{ccc} \text{YY} & \times & \text{y} \\ \text{Yellow Queen} & | & \text{Black Drone} \end{array}$$

Queen Yy Khaki	Queen Yy Khaki	× ·	Drone Y Yellow	Drone Y Yellow
Queen YY (f) Yellow	Queen Yy (f) Khaki	×	Drone y Black	Drone Y Yellow
Queen Yy (g) Khaki	Queen yy (g) Black		Drone Y Yellow	Drone . y Black

It can thus be seen that if yellow colour was by one gene, the daughters of a mis-mated yellow queen would all be " khaki " (or yellow if yellow was dominant). If a Khaki (Yy) daughter mates with a Yellow (Y) drone, her daughters (f) would be half of them yellow and half of them khaki. If she mates with a black (y) drone, her daughters (g) would be half of them khaki and half of them black. In either case her drones would be half of them yellow (Y) and half of them black (y), and never could a drone be khaki.

As any breeder knows, such definite colour results in such definite proportions are not obtainable, even after allowing for plural mating, and so it follows that colour in bees is emphatically the result of multiple genes. The number of genes involved is probably not very many because curious colour results are sometimes obtained especially with drones.

Supposing again colour was given by one gene, breeding would be as shown in the following diagram. *Aim :* to produce Yellow British Bees (the aim in view must always be clearcut when dealing with theoretical genetics). No colour dominance is assumed.

It is just a question of mating each generation of queens with pure-bred English drones, and rejecting the half which does not show khaki colour, and a little work at the end which is rather more tricky. It would only take a couple of years to do, and it is most unfortunate that as colour in bees is not governed by one gene it is all nonsense. Those who are interested in the subject can find an excellent diagram showing how the above system does work with chickens in *Animal Breeding* by Hagedoorn, p. 60. This only makes breeding bees still more difficult than it would otherwise be, and is still another reason why the amateur is advised not to breed " from his best stock " but to purchase a " breeder queen " from a reliable large scale commercial man.

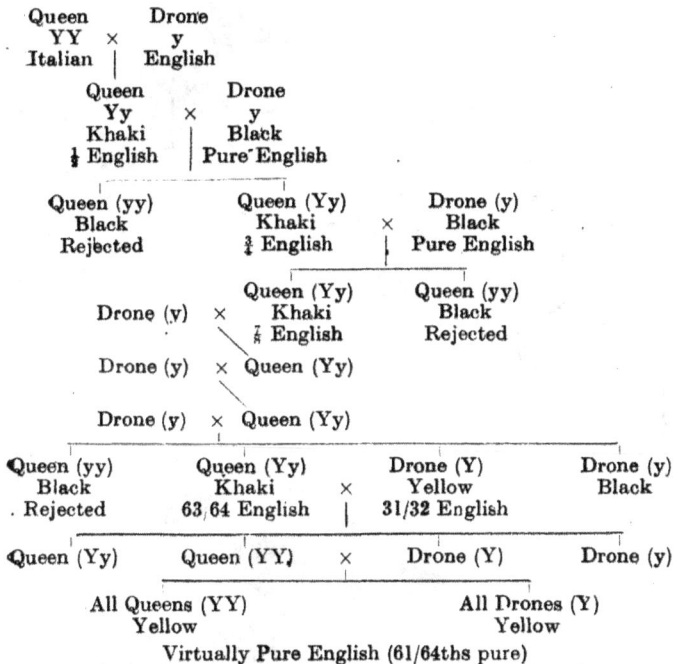

```
Queen      Drone
 YY    ×     y
Italian |  English

    Queen      Drone
     Yy    ×     y
    Khaki       Black
  ½ English | Pure English

   |_____|
Queen (yy)      Queen (Yy)        Drone (y)
 Black           Khaki       ×     Black
Rejected        ¾ English     |  Pure English

                   |_____|
                Queen (Yy)       Queen (yy)
 Drone (y)  ×    Khaki            Black
                 ⅞ English        Rejected

 Drone (y)  ×  Queen (Yy)

 Drone (y)  ×  Queen (Yy)

|_____|_____|_____|
Queen (yy)    Queen (Yy)      Drone (Y)        Drone (y)
 Black         Khaki      ×    Yellow           Black
Rejected     63/64 English |  31/32 English

|_____|_____|
Queen (Yy)    Queen (YY)    ×   Drone (Y)       Drone (y)

   All Queens (YY)              All Drones (Y)
      Yellow                       Yellow
       Virtually Pure English (61/64ths pure)
```

The cardinal rules are therefore to find a strain (not necessarily race) of bee which suits the scheme of management, size of broodchamber, etc., and to obtain a pure bred mated queen of that strain. Efforts must then be made to breed queens from her and from those of her daughters which have mated with drones of the same strain *in spite of the*

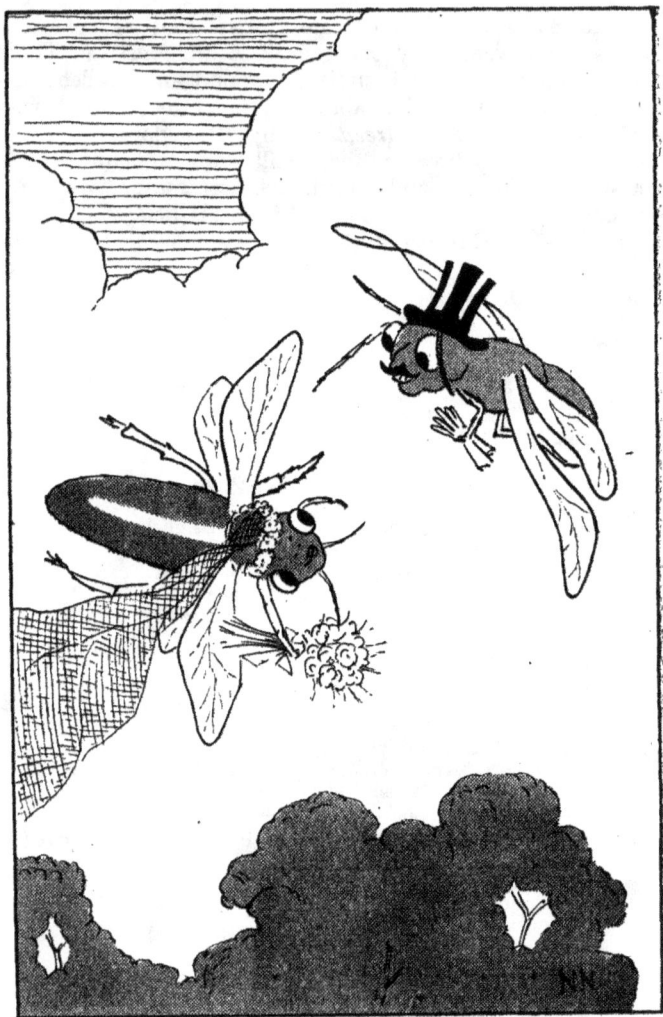

"Good show, Sandeman Allen's 10—1 chance coming off!"

fact that the mis-mated daughters may turn out to be better honey producers. Colour is another matter depending rather more on personal preferences and prejudices. So much skilful breeding has already taken place that yellow bees are no longer to be regarded as delicate and over prolific creatures suitable only for enormous hives in Southern climes. After all, American winters are more severe than ours, and most of the Americans whose livelihood depends on bees prefer yellow ones, the characteristics of which are far removed from those of their distant Italian (or should it be Cyprian?) ancestors.

The small scale bee-keeper has to realise that the matter of *establishing* a first class strain of bee is one that presents very real difficulties. The use of a breeder queen having an undesirable trait can easily wreck the results for a couple of years. The members of our local bee-keepers association were given a questionnaire, and among the answers regarding the crop was one which said " average crop 30 lbs. per hive, but if you discount the stocks of one particular strain the average was nearly 60 lbs. a hive." This strain had a local reputation as " good hardy strain of bees adapted to the district."

To start with, establishing a good strain in the apiary requires the selection of a breeder queen, and that is a matter which the small bee-keeper is most strongly advised to leave in the hands of a professional. Once she has been selected, the rest of the plan can be prepared and a way to start this is to requeen every single hive with the daughters of the breeder. It sounds simple enough, but quite a number of hives will still be found with the " wrong " queens in the autumn. Do not worry too much at this stage about the physical as opposed to genetical quality of the queens — the former can be attended to later. The following year, every drone in the apiary will be of the breeder queen's strain, the workers will be half and half. Any stocks which were not successfully re-queened may be examined once a fortnight and all drone brood " uncapped " with a knife, but if you do not carry out this drastic step, the chances are that no great harm will be done. When every drone in the apiary is of the required strain, the chances of mis-mating are fairly remote. We hear stories of drones flying miles in search of virgin queens and *vice versa*, but the chances of a virgin mating with a drone

of the stock from which she flies are said to be about five or
ten to one, and if she mates with another drone, that drone
normally comes from the same apiary. If it does not, it is a
most unusual occurrence. I am speaking of an apiary at least
a hundred yards clear of the next, and not of thoroughly congested districts which are, in any case, best avoided.

Thus, the first year is spent in requeening each hive in
order to get a uniform strain of drones. The second year is
spent in requeening each hive again, but, provided proper
records are kept, this is not such a desperate matter as the
first year's programme. In fact, the future programme is
rather negative and consists of ensuring that no virgin is
allowed to mate unless that virgin's mother is ether the original breeder queen or a fresh breeder queen, or is a daughter
of the breeder other than one of the bunch that were bred in
the first year. In other words, every virgin must be bred from
a queen which has " presumably " mated true, and that
rules out the first batch of virgins which mated with the old
strain.

I mentioned a " fresh " breeder queen. If the same breeder
is used in the second year, it comes about that " her " virgins
mate with what are virtually " her " drones. Almost at the
beginning of the programme the bee-keeper may start worrying about the closeness of in-breeding. An experienced bee-
keeper with a great many hives told me that the bees in two
hives had shown rather special qualities for many years. I
said that queens die, so surely these qualities get a bit ironed
out. The reply was that as those noticeable qualities had not
been ironed out, it must be presumed that the virgins always
mated with drones from their own hives. He prized these
stocks above others and as this close in-breeding over years
had in no way diminished their special qualities, you and I
need not worry. " What is the difference between in-breeding
and line-breeding ? " " It is this. If you go on breeding
from the same bunch of hens and after a bit you get left with
a rotten lot, you blame it on in-breeding. If, on the other
hand, the game is successful, that's line breeding." You can
do a certain amount of damage by in-breeding in a small
flock of hens in quite a few years. With bees it may occur.
Perhaps it does eventually, and perhaps it does not. It may
be that because they have fewer chromosomes it is of less

importance. In any case, it is a matter which need concern the small bee-keeper not at all. If the strain is not up to expectations, try a new breeder and don't worry any more about whether any deterioration was " caused " by inbreeding. Leave that worry to the know-alls. It is by no means certain that in-breeding of insects is harmful. In many types it is the rule rather than the exception.

It frequently happens at demonstrations and lectures that people ask, " Is all this queen raising business necessary ? " There is one good counter to that one, " Are you completely satisfied with your strain of bee ? " Quite often the answer is " Yes," and to me that is surprising. Ask any business man if he is completely satisfied with the article he manufactures, the land he farms, the service he performs, and the true answer is that he is not, and that he spends some time thinking of improvements or working to that end.

Now I come to a genetical point of interest. It is well agreed that a queen raised from a grub which has only been raised as a queen after it has reached a certain age is inferior. There are certain other factors, some generally accepted, and some arguable, concerning poor quality queens. But, is it better to use a poor quality queen of first rate pedigree, or a good quality queen of second rate pedigree, if these are the only queens available ? No doubt the honey farmer will plump for the good quality queen every time, provided the second rate pedigree is not what might be more aptly termed " third rate." But the small man, as usual, has a different problem because he cannot, normally, have a separate and secluded mating station. This means that, unless he is going to buy new breeder queens frequently and be rather clever with them, he is going to run the risk of deterioration of stock. *It does not matter how ill treated a pedigree queen may have been in any stage of her life, provided she has been fully mated and can lay eggs of both sexes satisfactorily, then her sons and daughters will be good, provided of course those daughters have been well cared for. This does* **not** *apply to animals. It may be possible that a queen bee could be so badly reared that her offspring could suffer in quality, but my own opinion is that the likelihood of this happening must be too remote to* *worry a small scale breeder, although I am open to correction.* Because I cull ruthlessly every queen of poor breeding, I know that the chances of a virgin from my hives mating with an undesirable drone are remote. Once the strain in the apiary is uniformly good, more attention can be given to the quality of the individual queens. That is a thing which may be delayed longer than planned and the reason is this. If you are going to have a queen rearing set-up, more queens will be produced than are required, because, (a) if you are going to graft a row of cells, you need a row of nucleus hives, (b) if you have a row of nucleus hives you need to keep them busy mating queens ; and (c) if you do this you will have too many queens. I took a number of orders for queens to use up the surplus, but I had not then heard of a remark by a competent large scale honey farmer to the effect that it takes a man at least three years to learn to raise queens. So all my best queens went to fulfil orders, and I had to keep those whose mating was overlong delayed. Even so, they have done me well and, of course, their drones are all first rate.

If you are going to raise queens " according to plan," you will need about eight mating hives and at least three double Nationals to stock them with and one (or two) building stocks. It is all rather an expensive performance. So, if you do not " want " to do it, leave it alone. I am writing for the man or woman who keeps bees for enjoyment as well as profit, so if queen rearing is not popular in itself, stick to one of the simpler methods.

At this stage, I would like to mention a point which has been laboured far too long, and that is the use of swarm cells. The usual argument is that if you use queens from swarm cells, you tend to propagate swarming tendencies in the strain. That, like most generalisations, is rubbish. If you always use swarm cells, it means that you only breed from stocks that swarm, and it follows that the " swarmier " the stock, the greater the likelihood of that strain being used year after year. The stocks that do not swarm, do not get bred from, and that would be a thoroughly bad way of providing for new queens. But, if you have a good breeder queen, and her stock decides to swarm, don't, don't, don't think that *those particular* cells are worse than cells raised *on that queen's* brood under one of the other impulses. Probably these cells will be better. This all looks so obvious and yet I have read " queens (produced) under this (swarming) impulse . . . . must have the swarming instinct developed to a higher degree . . . . " The author of this sentence did not enlighten us whether the swarming bees told the royal grub about it all, or whether they left a message for the virgin.

So, having decided which queen is the one whose daughters it is intended to use to stock the apiary, it remains to decide how those daughters are going to be got. There are many methods and it is quite easy to get muddled. Not long ago, the owner of about forty hives said to me, " I don't think I can be troubled with the method you use, and, in fact, I've started using the Manley method." Naturally, I was interested as anything new from that source would be worth investigating. However, it turned out to be the Miller method. It is so much better to be clear about the different methods than to be concerned with their inventors. I, myself, have invented a method of queen raising and anyone who wishes to christen it (politely) may do so !

A very simple method is as follows. The moment queen cells are found in more than one stock, " Snelgrove " the breeder stock or remove the queen and a small nucleus from it. This will provide you with queen cells of good quality and pedigree, provided you can go round on the evening of the 5th day and destroy the sealed cells. Now for the stocks that were building swarm or supersedure cells. Whenever your method of management calls for reducing the cells to one, or in any way arranging for the stock or a nucleus taken from it to have a virgin emerge, well, you have on hand in the Snelgroved stock just the cell for the job.

There are many ambitious methods of queen raising, but this is no place to discuss them all. Some methods call for cutting the combs, stabbing selected cells out of combs or otherwise mutilating combs. This may be quite worth while for very large scale men who ruin one comb and get a lot of cells from it, but it seems an awful waste to mess up a perfectly good new comb for half a dozen cells. I do not have to make my living from bees and I like raising queens as a hobby as much as for any other reason, so I shall probably continue to fiddle around with all sorts of methods as and when I have time, but even so, I find methods which call for cutting combs up offend my nature. Besides, methods involving cutting combs call for the use of newly built combs and these are especially valuable. So now ..I recommend three main systems, the grafting method, the method already described, and my own, as yet unchristened, idea.

The grafting method is well described in many books, so I shall only mention those points which are of special concern to the small scale queen rearer.

1. Making the cups. Soak the forming sticks in C O L D water.

2. Have the molten wax in a gluepot or some similar vessel with the receptacle for hot water under the wax pot.

3. A warm damp atmosphere for grafting. I wish I had a greenhouse. A handy dodge is to use a big, tinlined crate and tip a kettle full of boiling water into it. A few moments later, tip it on its side with the open side away from the wind, and get down on hands and knees and do your best " in " it.

4. It is so much easier to use a variation invented by Hillier. Place a comb of hatching eggs horizontally over the building stock and by the next day quite a number of grubs will have been selected by the bees and their cells will be flooded with royal jelly making it so much easier to get the transferring needle under them. It also gives you plenty of spare royal jelly to prime the cups with and to transfer with each grub. The jelly will be appropriate for the age of the grub.

5. It is rather fashionable to decry the use of wooden cell-cups, but for the small scale man, they are useful. When cells are ready for distribution, lift out the frame with one hand, grab a cell firmly by the wooden cup, twist it off, and back goes the frame. This, repeated for each cell, does away with the nuisance of cutting the cells off all at once, which really needs to be done in a warm place. Also, wooden cell cups fit into the holes in nursery cages so easily.

6. To fix the wooden cell cups on the bar, place some blobs of wax on the bar, touch each blob with a hot poker and straightway, press on the cell cup. The same hot poker can be pushed into a blob of wax inside the wooden cup to fasten the wax cup there.

7. Transferring needle. The home made ones are normally better than those which are bought. Hammer the end of a bit of thin wire into shape, clean off jagged bits and fix the wire into an old penholder.

8. Double grafting is a proceeding sometimes frowned upon, but really is not at all bad. If you graft a row of cells on Saturday, it may be too soon after " preparing " the stock. A few cells may be accepted and can provide royal jelly for priming the others. Also the bees will have " worked on " the cells. The queens resulting from double grafting will be no better than those from ordinary grafting, but, in amateur hands, there will usually be more of them !

9. The bottom bar of a shallow frame is a simple and satisfactory place to affix the cell cups. Paint the top bar of this frame so it is obvious which one it is.

Now, I come to my own idea which will require more space because detailed instructions cannot be found elsewhere.

I want to emphasise that this method is only in an experimental stage. I hope that other people will try experimenting with it and will improve on it. For the present I hope it will be known as my queen raising idea and not my queen raising method. I would rather publish my idea, knowing that it can work well, than spend years experimenting with it when others better qualified than I could probably improve on it more quickly.

The cell building stock must be very strong indeed, stronger than cell building stocks for other methods. It can be a stock which is found to be queenless and containing sealed cells. Such a stock will have swarmed and the swarm will have returned during the week, because the clipped queen will have been unable to get away. When using a stock with queen cells (all of which must be removed), it is as well to give some unsealed brood the day before, to get the nurse bees' glands working again, but remember to remove it. Weekend bee-keepers should then place the prepared frame in position on Sunday, ready for the following Saturday week.

The cell building stock must have a broodchamber, queen excluder or two, and super, then an empty super, crate or eke, then the crownboard and beetight roof. So there is an ordinary supered hive with a space between the top super, frames and the crownboard. The comb on which the cells are to be built should be laid flat across the super frames, supported on an empty frame or something to give room for the cells to hang down from the face of the comb.

The frame to be used is best if freshly drawn and filled with an even mass of grubs which have just hatched, grubs which are just hatching, and eggs on the point of hatch. I find I have an advantage here, because my breeder queen is in a National and my cell building stock is a Dadant, so there is plenty of room to put a B.S. frame flat in the top of the hive.

The bees will raise a number of cells which will hang down from the face of the comb. Being new comb, the bees will model the cells easily. The queens will have been fed as queens from hatching or from a few hours afterwards, and no upheaval will have taken place such as grafting, etc., so I hope the purists will find no fault with the system.

On the thirteenth day, which, for weekend bee-keepers will be the following Saturday week, visit the hive with a good collection of queen cages. Nursery cages are good, but keep them in a warm place and not out on top of the next hive roof for the wind to whistle through. If all has gone according to plan, one virgin will have emerged and the bees will be attempting to swarm with her. She will be trying to find a way through the excluder to join the swarm which is why two excluders are a wise precaution. Meanwhile, bees will be preventing the other virgins from emerging, and will be feeding them while penned up in their cells. The tongues of the imprisoned virgins will be visible sticking out near the ends of the cells. This frame can be removed from the hive and placed flat with the cells pointing up. With a matchstick or even fingernail, prize off the cap of the cell with the most " active looking " tongue sticking out of it, and the virgin will crawl out and go racing over the comb face. Catch her before she gets away, pop her into a nursery cage or whatever you use and get on with the next cell. The first time I tried this system, I had a dozen cages filled in five minutes and then started casting round for empty match boxes.

When all the virgins have been liberated from their cells,

go through the super of the building stock and hunt up the virgin that will be loose in it. Alternatively, remove the excluders and she will go down on her own legs instead of being " put." Better still, remove her and cage a mated queen in the brood nest.

This system is really the most simple system imaginable for producing a dozen or so virgins. No special equipment is needed except an empty super or crate and a number of empty matchboxes. Even the comb, when finished with for queen raising, is not damaged or spoiled and can be placed in the brood nest of an ordinary hive to let the worker brood emerge. The building stock can be just an " awkward " lot which will not respond to swarm control treatment, so nothing is wasted.

But it has its limitations. No doubt it will be described as a system which may be all right for the small scale man, but for some doubtless admirable reason, will be unsuitable for the commercial honey producer. Also, it is not so good for the man who only requires two or three virgins which can more easily be produced in a divided box over a Snelgrove board. But, apart from these, there is a more serious objection, or rather limitation. The success of the system *entirely depends upon the intention of the bees to swarm with the first virgin which emerges.* So, if the bees decide not to swarm, the first virgin out will destroy all the others. This means that it is effective during the swarming season *only.* Also, it is as well to prepare another stock a week later, so if the first happens to fail, there will be another chance next week. Even, by the way, if a stock does fail to produce more than the one virgin, if it was a swarming stock in the first place, you haven't " wasted " much because all you have done is to " requeen with a virgin after a break in brood rearing " which is an accepted method of swarm control. Do not rely too much on this system—when it works, it works wonderfully, but failures are frequent. It needs more experimenting with. It would probably be better to place the building frame, when the cells are sealed, on a stock which is actually ready to swarm of its own accord.

Whatever method of queen raising is used, do not start it until drone cells have been sealed for *at least* 8 days, otherwise the first virgins will find no potent drones.

Queen Introduction

Beyond the fact that I often use a cage of my own making, I have nothing much new to say. Snelgrove has written a most interesting book entirely devoted to the subject.

I have hardly any experience of direct introduction. Getting queens raised, introduced to nuclei and mated, is no end of a lot of trouble. When I finally see the mated queen in a nucleus, I still regard her as an achievement and as something valuable, so I can never quite bring myself to put her direct into a stock, " unwrapped " so to speak. Most of the instructions for direct introduction end with a sentence like this : '' go to the stock in the evening as quietly as possible and with as little disturbance as possible, and run (!) the queen under the covers or through the feed hole with a puff of cigarette smoke as a chaser.'' I was getting a stock ready one Saturday evening after I had steeled my heart when I realized the excluder must come out. Something wrong no doubt, but I removed it (disturbance). An hour later when flying had stopped, I lifted the lid and blew cigarette smoke into the feed hole and in went the queen on to the top bars of three supers. Strangely enough she was still alive when rescued about twenty seconds later, for all the world like a dud penny out of the bottom of a slot machine. It seems clear however, that direct introduction is, in the hand of practised men, as sound and safe a method as any other, and, I understand, far less trouble.

To the weekend bee-keeper, I would say it is best to read up this subject in modern books and journals, and get a grasp of the conditions in the stock which make introduction favourable. The condition of the receiving stock is a long and controversial topic and is probably more important than the type of, or lack of, cage.

One last point about introduction. The usual advice given is not to open a stock for a week after introducing a queen, and as a general rule, it is sound. The idea has taken root in the minds of many bee-keepers that the new queen is " on probation '' and that the bees '' realize she is new and take a week before they are content to have her.'' I think that must be nonsense. Bees just haven't got that much grey matter. If you remove a queen in full lay and substitute another in full lay, either she is fully accepted about as soon as the bees have sorted themselves out, or she is not, and is out. If, however, you remove a queen in full lay, and substitute one that is not, obviously a sense of " something wrong " will be felt in the hive—the normal rythm will be upset. In the case of a travelled queen, there may be a two day break in hatching ; in fact, it is not until eight days after the queen starts regular laying that the rhythm is normal again. If you put a queen laying at normal rate into a stock as a simple replacement, and very little candy is put into the queen cage or if direct introduction is used, it is normally safe to open up two days later and probably next day if there is a honey flow on.

Cages

I use two sorts. The Manley type cage, which is illustrated in his books and can be bought from appliance dealers, is probably the best, but read his comments on its use. The other cage I use is easily made. The plain cage, sometimes called semi-direct, most bee-keepers will know. It is just two small cubes of wood, one at each end of a wire mesh tube. You put the queen in through a hole in one cube and stopper it.

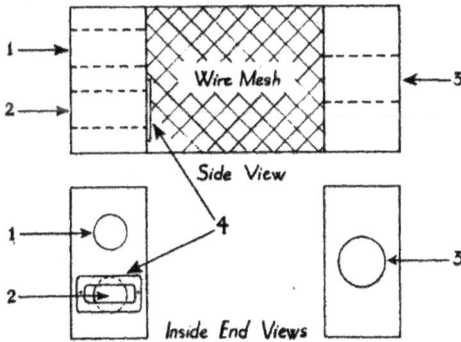

Queen Introducing Cage

1 Exit hole for queen, filled with candy.
2 Entrance for workers, half filled with candy.
3 Hole for inserting queen in cage.
4 Slot of excluder zinc.

The bees release the queen by eating out the candy from the hole in the other cube. I make them up with larger pieces of wood at each end of the wire mesh tube, and put two holes side by side. One exit hole has a piece of excluder across it and less candy is put in that hole ; simple Chantry. I think it is slightly more efficient than the simple cage, and very little more trouble to make. In a nucleus hive, this cage can be laid flat across the top bars, which saves a bit of trouble.

If you are making cages, make the exit holes $\frac{3}{8}''$ in diameter and run a hot poker through them to burn out the splinters. A $\frac{1}{4}''$ hole is big enough to let a queen through, but as I have had queens dead in exit holes, I prefer them on the large side.

Queen Cage Candy

I am only mentioning this because, in spite of the fact that practically every book has the recipe in it, quite a lot of bee-men I have met still think the ordinary candy will do. It will not. Queen cage candy is simply a paste of liquid honey and icing sugar, and if you are making it for the first time, start with a very little honey to quite a lot of sugar, or you will run out of sugar !

CHAPTER VII

THE HARVEST

I HAVE yet to find the book which really gets down to the job of telling me how I should have tackled the many problems I have come up against on extracting day. Quite a lot has been written about the methods used in extracting tons of honey, but books for the small scale bee-keeper usually content themseves with such statements as :—'' The frames are uncapped (described) and placed in a machine called an extractor wherein they get whirled round until all the honey has come out.'' It all sounds delightfully simple, so I will describe my own experience.

When I started keeping bees I was given a little extractor which held three B.S. shallow frames, and it worked quite well. My troubles really began when I started using Dadant hives. At that time a small extractor to take Dadant frames was not listed in any catalogues I saw. In one catalogue there was a machine advertised as holding six B.S. shallow frames and another model described as being the same type but capable of holding all frames up to M.D. deeps. Yes, it was big enough to hold a Dadant brood frame in each pocket but not big enough for two shallows. It was made half an inch too small ! For one thing this meant paying for a six frame machine which only held three frames, and for another, the delivery date was described as '' uncertain.'' By mid-summer I had only one Dadant hive bringing in honey and about ten others containing two frame nuclei. I could not buy Dadant shallow frames, because the post war manu-facture of M.D. equipment had not got into its stride, so I supered with brood bodies until the hive stood five brood-chambers high with well over two hundredweights of honey in it. I ordered a radial extractor ; it was the only thing to do. Everything about it was vast, even the crate it came in had a £5 deposit on it. Then came the shocks. First it was too heavy for me to lift, and then it was too big to get through the front door. Luckily I could get it through the

back door and into the kitchen, but further into the house
it would not go, and it was obvious that it could not sit
about indefinitely, occupying half the kitchen. The sooner
extracting was done the better, and the extractor taken away
by horse and cart to be stowed in a neighbour's hayloft.
Then there were more troubles, I had read that uncappng
could be done into a basin lined with muslin, and when
finished the four corners of the muslin could be gathered
together, hung up, and allowed to drain. By the time I had
finished I had quite a lot of cappings hanging about, drip-
ping into pudding basins and suchlike. They dripped for
days and I expected to be denied the use of the domestic
pudding basins the following year. But to return to the extractor. I had to set about the rotor with a hacksaw before
it would fit Manley type shallow frames because the slots
were too small for the wide sidebars, but that was of course
the following year.   The solution is to have the slots cut out
like

so that Manley type frames can '' house '' into the wide part,
and other frames right down into the narrow slots. This
extractor gave me no end of trouble. It went back to the
makers who reported that there was evidence that it had
been turned upside down and jolted while in transit, and the
damage slowed down the rate of turning so much that,
although it was no fault of the machine, I began to have
quite a real dislike for it. But in the end I was sorry to see
it go because before it went it proved to be an efficient
machine. It held 21 frames at a loading, and ten minutes
hard turning got the frames pretty well clear of honey. It
really needed a permanent stand in an extracting shed,
where it is now. If I still had it—and the shed ! — I should
bed down its stand in cement and bed down a second hand
bicycle near it. I am sure a little ingenuity and a long
bicycle chain would minimise the most disagreeable feature
of these hand powered radials, namely the physical effort
required to keep them running so fast and so long. If I
could pedal the thing I am sure it would be easier. No
doubt an electric motor would be the thing to use, but

" The extractor had to be stored by a neighbour."

makers' catalogues always show electrically driven models
at a vastly higher price and electricity has not yet got this
far into the country.

Apart from the inconvenience of keeping the extractor in
a neighbour's hayloft, and hiring a horse and cart twice
whenever I required it, there was another reason why I made
up my mind to have a new machine. A little machine could
be stored in the attic, and in *Bee-Keeping in Britain*
Manley had written "A ton of honey can easily be extracted by means of a two-comb tangential extractor and a
couple of small settling tanks." Like everything else in that
book, this statement is correct, but I made the mistake of
assuming that everything was suitable for and would fit in
with my particular arrangements for extracting. In any case
I doubt if it is good general advice for serious *weekend* bee-
keepers.

I decided to have a six frame machine, or rather it was
decided for me because as far as I knew no other sized
tangential machine were made. I had already done most of
my extracting with the big radial before I sold it, and all
that remained was about thirty frames of unsealed honey
rejected at extracting time and put back for the bees to finish
off. There was also a super of newly drawn B.S. Hoffman
deep frames. My new extractor had the usual wire mesh
grids of the tangential machines, and I noticed what looked
like a serious snag when I put in some empty frames to see
if they fitted. The wide sidebars of the Manley type frame
contacted the grid but the face of the comb did not. I put in
a Hoffman B.S. frame and it was the same, the " shoulders "
of the Hoffman frame held the comb face a quarter of an
inch off the grid. I wrote to the makers who told me I need
have no anxiety. Well, every frame I tried in it was smashed.
The B.S. frames did not reach the top and there was nothing
to stop them falling about in the extractor, and they broke
too. The manufacturers went to no end of trouble in the
matter and designed special grids, but these, although they
reduced the incidence of breakages, brought other disadvantages in train. Also, it seemed to me that experiments with
other grids were just a palliative—what was needed was a
drastic overhaul of the design to keep abreast with modern
frame developments. Until some manufacturer has done

something about it I would recommend those using any sort
or size of Hoffman frame to use a radial. The Americans
may have a solution, and from a careful examination of the
illustrations in an American catalogue I think they have,
but it is an easier problem for them as they only cater for
Langstroth and Dadant frames which are the same length.
B.S., 16 × 10, and Langstroth frames when placed in a
tangential extractor have the upper sidebars at different
levels ; it must be heartbreaking to be a supplier of appli-
ances ! The firm was good enough to take back my extractor, for which I am grateful.

I believe frames stand up to extracting better if they have
been bred in once. It is also possible to extract any comb in
a tangential extractor without grids if you go slowly enough,
and keep reversing the frames until they are practically
empty. This is especially so with Lee's Renstan which holds
the frames at an angle, that is part way between the radial
position and the tangential position. In any case, either you
break combs or you go terribly slowly and take a lot more
time if you can spare it.

But quite apart from the troubles I had with the tangential
machine, I can now see, looking back on it, that it would
have been too small for me. The problems facing the weekend bee-keeper are really quite different from those of a
small scale professional. " Two small settling tanks " is the
key to the difficulty. This means two hundredweights each
weekend—start in September and if it is a really good year
you might be through by Christmas, or you might not ! My
own plan is to get a load of supers into the house and all
equipment, extractor, tanks, etc., ready and washed on
Saturday. Extracting is started with an assistant to help on
Sunday morning, and everything is planned on the assumption that about 7 cwts. will be extracted and all extracting
equipment washed up and put away and all supers returned
to the hives for cleaning *that day*. Bottling or canning can
be done in the evenings after my office work is done, but
extracting is a tough job ; two weekends and it should all
be finished even if there is a ton to do, although in that case
it is less burdensome to spin it out over three weekends.
*Now it is obvious that the serious minded weekend bee-
keeper has either got to have equipment almost, one might*

*say, out of proportion to the size of his undertaking, or
he has got to do without his holiday.*

In the end I bought a Parallel Radial from Mountain Grey
Apiaries. It cost me over thirty pounds. With it I can extract half a ton a day. Not only does it satisfy all my ordinary requirements, but, being long and narrow as opposed to
round, it can be carried through the house and stored up
against the wall in my little work room. It extracts 21 frames
at a time and about five minutes turning is enough.

That is the story of my own trials and troubles with extractors, and I hope that by telling it in that way people will

get a better idea of exractors than from a summary of the advantages and disadvantages of the different makes of extractor.

There are various points to emphasise in the above story. If you are going to keep a score or so of hives, you must be prepared to deal with a ton of honey whenever a really good year comes along. I think it is in any case asking rather a lot of the human frame to deal with a ton of honey and a full time job at the same time, and if it is going to be done you will, apart from having the right equipment, need to get things organised so that the work will proceed smoothly and without hitches or bottle-necks. I hire an assistant on extracting days, partly because my back will not stand the strain. Even a boy is a help as he can pass the frames to the uncapper after cleaning the propolis and brace comb off the woodwork, transfer the uncapped frames to the extractor, remove the empty frames from the extractor and generally make himself useful.

I do not like the word "bottlenecks," but nowadays all of us know what it means. As I have said, extracting day needs a lot of careful planning if the work is to proceed smoothly, and there are three main places where bottlenecks and other troubles may be expected—uncapping, extracting, and getting the honey into the tank. The extractor as such has already been dealt with, but I must say something which has been said before, notably by Manley, about taps. When I had my big radial I used to open the tap when I started, but the speed of extraction was such that although the tap was open all the time, the honey level rose to the rotor and held up work. It just would not flow through the little tap fast enough. I mentioned my grievance to the makers and their representative told me that these machines were exported and that the firm had to study the needs of the man overseas who appeared perfectly content with the machine as it was and would probably not be content if the price was increased by the additional cost of fitting a larger tap. It seemed that made some sort of sense to him. Different firms do use different sized taps so it is as well to check up before you buy. I¼ inch is the *smallest* acceptable size for a tap fitted to a radial, and three inch taps are far more satisfactory.

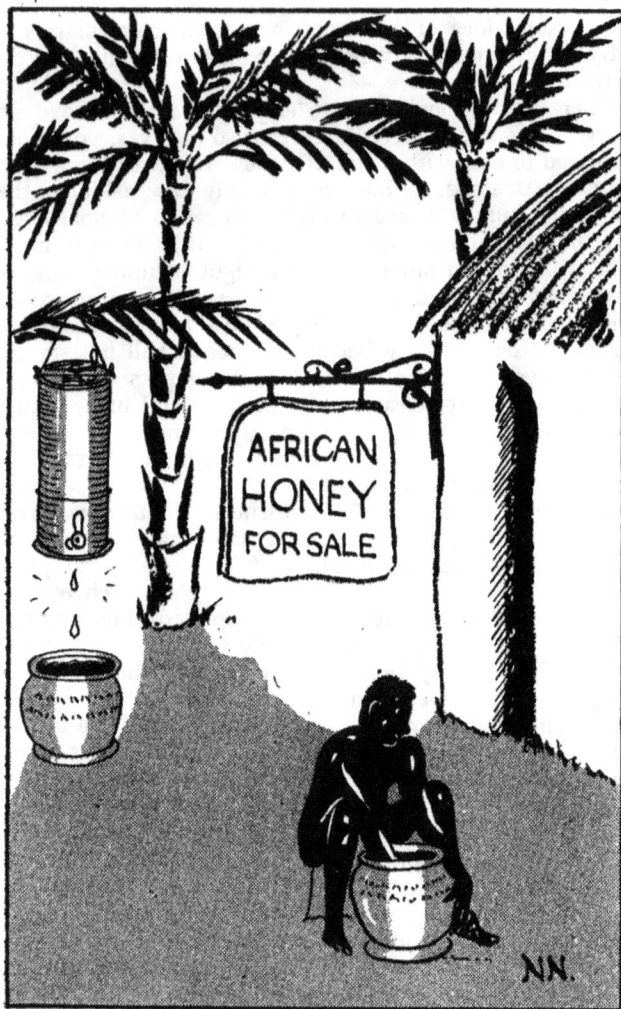

" The firm had to study the needs of the man overseas,
who appeared perfectly content."

Storage Tanks and Ripeners

In this country it seems that if a honey storage tank is cylindrical and tall it is called a ripener. Any other distinction or nomenclature is purely historical except among the pundits of the old school of bee-keeing.

There are certain special advantages which attach to each of the two shapes, and here as usual I am treating the matter from the point of view of the weekend bee-keeper. A 40 gallon galvanised iron tank holds over 6 cwts., can be got from any builders merchants and costs about £2. The only difficulty is fitting the tap. If the usual lift-and-drop gate-type is used it should be fitted on to a cone, and for some reason connected with metalworking, tin cones are difficult to fit to galvanised iron tanks. I have a spring loaded tap on my tank as well as a couple of those tiny turn-button taps, all of which have back nuts and are easy to fit. These little taps are cheap and can be fitted four inches apart so that two can either flow into two or into one 28 lb. tin while two others can be used to fill jars. Treacle taps are fairly expensive. A 2-inch tap or bigger would save a lot of time filling 28 lb. tins, but I do not know of one which is both inexpensive and easy to fit. (Since writing this, Messrs. Robert Lee, Ltd., have provided me with a sample tap of a fair size which can easily be fitted on to a galvanised iron tank). A tinsmith can fit a " spout " to a tank without much trouble, and if it is only to be used for 28 lb. tins, it can be opened and shut by using a bung. The weekend bee-keeper must decide what sized tanks to have in quite a different way from the small scale professional or anyone with a score of hives and plenty of spare time.

A really good harvest from a couple of dozen hives may be as much as a ton (and I, who say it, do not come from Gretna Green). How many weekends are to be set aside for extracting ? If the answer is " normally two but three in a very good year," then estimate the crop expected in a better than average year and get tanks to hold half that quantity. I keep two dozen hives. In a good but not exceptional year the yield would be twelve to fourteen hundredweights, so I have tanks enough to hold seven cwts., and anything over that quantity either means three weekends or else I borrow ripeners from friends.

One large or two smaller tanks are necessary to hold the bulk of the honey as extracted for a couple of days at least, until the froth rises. The honey is then run off, and the froth can then be collected, warmed, and put into a small ripener when still more honey can be run off. For this purpose a little ripener, tall and thin, is excellent, and it has other uses too. Mine holds the contents of three 28 lb. tins which is a convenient amount to bottle in an evening for winter and spring orders.

There is of course no need for the main honey tank to be square. A couple of nice new galvanised iron dustbins holding 2½ cwts. each can be fitted with taps and are cheap and light. Also, if they get a bit " tatty " they can revert to the use for which they were manufactured. They have nice lids to keep out the dust. If the inside appearance is not all you might expect it to be, you can put on a coat of aluminium paint.

I advise using large tanks and running all the honey from them into tins. One small ripener can be kept for bottling.

It always amazes me that there are so many who keep up to a dozen hives, and who do not possess a ripener in any form at all. " Get it all over and done with at once " is their cry, and a conical bit of perforated zinc is suspended from the extractor tap. In between spins an assistant (usually a long suffering wife) opens the tap, the honey flows into the cone of zinc, " strains " through it, and is finally " caught " in and down the sides of the jars. A little honey tap is very cheap and it is the work of a few minutes to fit it to a 56 lb. honey tin, so about 7/6 buys an effective little ripener from which bottling can be done with very little trouble and with no mess at all. " But that," I have been told, " means extracting one day, bottling the next and honey in the kitchen for two days." Such people have for so long regarded bottling as a messy job that no other idea can germinate in their minds. The place for the honey is inside the jars, not down the sides of them, so if the kitchen is out of bounds on D Day + 1, take the ripener on to the dining room table. And please, if you bottle direct from the extractor, don't write to the Hon. Secretary of the Local Association and suggest he should " Do something to boost the sales of English honey." Buyers do not like froth on the top. If you do not wish to buy a ripener on the grounds that you have enough gadgets already, run off the honey into pails or tins, clean up the extractor, remove the cages and pour the honey back into it. Bottle the honey a couple of days later without using a tap-strainer.

If you are going to extract a hundredweight or two per day, a small melter or some other system is quite adequate, but if you only have weekends you will have to get a move on. After uncapping fast for an hour into a little melter I found the wax did not melt nearly fast enough and began to pile up on the melter. Perhaps a lot of wax cooled the small surface of the melter, but anyway it created a bottle-neck and the extractor lay idle while I hung about and watched the wax slowly melt away. If you coax or chivvy it towards the outfall it fails to set in a lump in the receptacle put to catch it. A big melter costs more, I regret to say, but not much more.

There are special receptacles made for catching the honey and wax from the extractor, and these drain off the honey from below the wax by means of a bent pipe. Manley dislikes this system because the honey is kept hot longer that way. He recommends using pails and putting these aside to cool and then lifting off the cake of wax. It is a good plan but stand by with at least four pails or you will burn your fingers trying to pick hot wax off hot honey ! You can fit a little tap to the bottom of a 28 lb. honey tin and just run the honey off from time to time. One way and another quite a lot of pails seem to be pressed into service on extracting day.

Uncapping

Here especially the weekend bee-keeper is up against it properly. There are instructions galore to be found in every book on bee-keeping, but by no means all of them are really suitable for the weekend bee-keeper. The disposal of cappings can take up a lot of time on extracting weekend, and on that day there just is not any time to spare.

One quite useful method is to take a tank (one of those dustbins will do admirably), fix a bar across the top with a spike or a slot in the middle, and steady the frame on the bar when uncapping. When uncapping is finished, stir the

whole lot up and leave until the next day when the honey
can be run off from the bottom. There is not even any need
to fit a tap, a hole with a cork in it will do, unless you have
any children who are likely to get at the cork.

Another method is to get one of those old fashioned wash-
hand-stands which look like a table with a big hole in the
middle to pop the basin in, and a shelf near the ground.
Sling a thing like a military kit bag but not so deep under
the hole by attaching it to a circle of cup hooks screwed to
the underside. Uncap into the bag and the honey drains
into a basin put on the lower shelf. If the bag is of muslin it
may burst, and if it is too substantial it will retain the
honey, but a muslin bag will do if a wire waste paper basket
is hooked up outside it. Linen cheese cloth is strong enough
in itself, but you need a big bag and you may have trouble
at the places where it is supported, or at the seams if any.

I dislike both these methods. As I have said before, most
weekend bee-keepers have to " tidy up " by Sunday night,
and I used to find that sticky bags of wet cappings were
never finally attended to until after Autumn feeding.
Bottling and feeding have to take precedence over cappings.
A cappings melter was the obvious solution, and has the
added great advantage that the uncapping knives can occupy
the trough of boiling water at the end of the melter. With
only a couple of dozen hives I thought a small melter
(Pratley uncapping tray) would do. It most certainly did
not " do."

Straining

If no straining is done, a carpet of froth, beeswax and
pieces of bees floats on top of the honey in the tank. I believe this can be " skimmed off " but I have never discovered how this is supposed to be done ; no ordinary kitchen
utensil seems to cope with it. A very simple straining device
can be made out of a length of mutton muslin which is
bought in the form of a long sleeve. Tie a knot in one end
and arrange for the other end to be open and at the top of
the tank. There are many ways of doing that, for example
tie the top round the kitchen steamer. A cylinder of wire
netting or something to stop the bag bursting is necessary.

There are many arrangements that a little ingenuity can

devise. I wish I had an O.A.C. strainer but there are limits
to spending, so I make do with some home made contri-
vance which usually gets forgotten about until the morning
of extracting day, when it is too late to construct it with
care. It is one of the first things which " goes wrong."
(Finishing time is always a few hours later than planned).

Bottling

It is as well to leave the honey in the tank for three
days unless it is fairly warm. That thin layer of foam on
the top does detract from the appearance of a jar. There are
good reasons for not bottling the lot direct from the tank, the
most important being uneven granulation, frosting and set-
ting very hard, all of which are quite likely to happen to each
pot. Far better to bottle what is needed and put the rest in
tins. Every now and again through the winter and spring
these tins can be warmed through, tipped into a small ripener
and bottled. It has often been asserted that the jars should
be " tipped " when being filled, by which is meant the stream
of honey should not fall slap down in the middle of the jar,
but should be managed after the manner in which beer is
poured out of a bottle into a tankard, and, I suppose, for
the same reason. This is nonsense. If there is any froth, then
unlike beer, it was there beforehand and no amount of
tipping will remove it. Any big bubbles trapped in the
honey will rise to the top, burst, disappear and need concern us no more. It has been shown that frosting may start
where the stream of honey strikes the glass. It may one day
be shown to be desirable to tilt the bottle to deter frosting,
or it may be the other way round. I have had practically no
experience of frosting. The tops should be well screwed
down to keep out air and moisture.

Warming Honey

As I have just said it is best to bottle the bulk of the crop
during the winter. It is easy enough to say " warm the honey
through and bottle it " but it is not nearly so easy as it
sounds. I have seen special devices advertised for warming
tins of honey but the price was very high. As the greatest
danger is in overheating, the important thing is to bring
the honey up to a temperature of about 125° as slowly as

possible, and this slow rise in temperature is most important at the early temperature ranges, i.e. at the beginning of the heating. Most houses have a hot airing cupboard and it is best to put the honey to be warmed into that the day before warming proper is commenced. If only one tin is to be done it can be suspended in a copper over a very low gas adjusted to bring the temperature up to about 125° after about 4 hours. If it is required to heat up several tins my advice is to put the honey into the airing cupboard first and then into a good sized strongly built wooden packing case electrically heated. The electrical heating unit may be 100 watt bulbs or small portable units, and in the construction of this heating cabinet my earnest advice is to consult somebody with a knowledge of electricity.

Labelling

Appliance dealers will supply booklets of various colourful labels on which your name and address can be printed. Provided you only want letters as opposed to pictures, and provided you buy about a thousand at a time, it is cheaper to design your own and have them printed. My only suggestion is that an oval label is far easier to stick on than a square one and, if not too deep, will fit on a ½ lb. jar. You can put the paste on with a brush, but I usually find the paste gets on the wrong side of the label, so I prefer to smear the paste on to a bit of cardboard and rub the ends of each in the paste just long enough to leave a sticky trace and put them on the bottles one at a time. I do not think a picturesque label helps to sell honey, but I am convinced that including the name of the county does help. There is no copyright about the label I use, so here it is :—

No. 588
EAST SUFFOLK
BEE-KEEPERS ASSOCIATION
½-lb. Net.

Finest Suffolk Honey

From the Apiary of
A. L. Sandeman Allen,
All Saints, Halesworth

Marketing

As Secretary to a Bee-Keeping Association, I am always being asked to "Do something about a marketing scheme." I regret to say I am not helpful. So far, I have had no great difficulty in selling my own honey and I do not feel disposed to sell everyone else's as well. I find some shopkeepers are most unwilling to take ½ lb. pots and it is curious that in more than one case I have almost forced it on them, only to be told a fortnight later that it was all sold, but they would rather have only 1 lb. jars in future. Those shops which sell both sizes in this district sell roughly equal numbers of each.

The weekend bee-keeper has a very real advantage over the large scale producer in this matter of sales. A far greater *proportion* of the crop can be sold retail, given away as presents, or just eaten. I think I am unusual in that I have more than once run out of honey through complying with orders and gone and bought 28 lb. tins from a commercial producer so that I can have some for myself. Others may buy honey for themselves, but I have never heard of it. I suppose the chief reason why I started bee-keeping was because, at the end of the war, I could not get anything like the quantity I should have liked. I set a good example !

Quite a lot of my crop is sold in 7 lb. tins. I charge for the honey and charge for the tin. When the tin comes back for refilling, it is not, of course, charged for again, and a number of tins do come back. There are all sorts of tins, those with lever lids and those with deep lids. I do not find the ordinary lever lids are safe in the post or by rail (nowadays), unless the honey is well granulated. Tins with deep lids are fairly safe in a box. The cardboard boxes chemists throw away when emptied of tinned baby foods just fit 7 lb. tins. Practically all the honey I sell privately goes in tins, and practically all sold through shops is in jars. I think the person who buys in a shop likes to *see* the honey, while those who get it from me direct get talked into having a tin ; a tin is usually a bigger order and it is so much less trouble.

Wax

For every 100 lbs. of honey, you get about 2 lbs. of wax, i.e. about 20 lbs. per half ton of honey. If you take very

little trouble, you will get 18 lbs., and if you take a lot of
trouble, perhaps 20 lbs. So, for 2 lbs. of wax, it is not worth
using up a lot of gas and electricity boiling it up and getting
a lot of utensils messy. In other words, wax is not one of the
things a small scale man need get unduly fussed about on
account of its value, but the time factor on extracting day
is a matter which needs thought.

Most books describe the harvest of wax as a messy
operation and the authors appear to write with feeling. I
have a theory about wax and mess, and it is that if all wax
is in cakes *before* rendering day, this day can be postponed
until winter and should be less palaver. A solar extractor is
not difficult to improvise, especially if you can find a box to
start with. The local carpenter will frame two sheets of glass
a quarter of an inch apart and fit them with a hinge on the
box. Choose a box big enough to hold a frame, and preferably big enough for a queen excluder. A piece of corrugated
galvanised iron and a trough completes the thing, and detailed instructions for construction can be found in the more
comprehensive manuals. Of course, you can make the lid
and all yourself if you are clever, but it is not easy to make
a lasting lid holding two sheets of glass fitted accurately
and designed to be out in all weathers. Remember :

Lord Finchley tried to mend the electric light
Himself, It struck him dead ; and serve him right !
It is the duty of the wealthy man
To give employment to the artisan.

Don't forget that the top sheet of glass must be held in
place at the lower end with a screw or knob so that the rain
can run off instead of being caught in the framework.

All odd wax goes into this and comes out in lumps, and
cappings wax comes off the melter in lumps. These lumps
can be put away and dealt with in the winter. Now a word
of warning :—If you want peace in the home, don't use
kitchen saucepans and other utensils for rendering wax !
They are the devil of a job to get clean again afterwards.
The weekend bee-keeper can make do with " small man's
equipment " here which is rather a relief. There are various
patterns of wax extractors in catalogues which can be used,
or the wax can be boiled up with (rain) water in a big sauce-

pan reserved for wax. The M.G. Wax Extractor is the one I use, and I find it efficient. The only nuisance to me is standing about "watching for the wax to boil." No kettle ever took so long, and wax boils over just as furiously as milk, and not only is it dangerously inflammable, but it makes a horrid mess of the cooker.

Wax moulds. If you want to show your wax, there is an excellent book which will tell you just how to set about it. If you do not, then the wax goes to make foundation and the makers are not concerned with the shape, and cracks in it. The simplest mould is made by cutting the bottom out of a 14 or 28 lb. honey tin with a wheel type tin opener. Put the lid on as usual, set the tin upside down and run the wax in through the bottom as it were. When quite cool, remove the lid and push the cake of wax out.

CHAPTER VIII

PAPERWORK

Income Tax

I HAVE several times been asked whether the profits from bee-keeping are liable to Income Tax. Although I am qualified to write on the subject, what I say is, to use official phraseology, "not necessarily the opinion of the Board of Inland Revenue." Besides, individual Inspectors of Taxes may have different ideas and the circumstances of no two bee-keepers are alike. However, a few hints may be useful.

In the first place, the Inspector of Taxes is not out to levy tax upon "hobbies" whether they are alleged to be profitable or unprofitable. The crux of the matter is, there-fore ; " it depends what you mean by hobbies," and the answer is entirely dependent upon the surrounding circum-stances. I have known three hives regarded as a business and taxable venture, and I have known twenty-five hives regarded as a hobby.

One thing is fairly certain ; if you buy equipment whole-sale and sell it retail, that points to a taxable profit.

If you are a fruit or seed grower and bee-keeping is part of your farming operations, then the profits (or losses) are normally taxable.

If you have special ".Bee " stationery printed for bee " business " letters, it points to a business venture. If you keep a large number of hives ostensibly for profit, that too points the same way.

The difficulty is to fix a line beyond which the venture may be regarded as taxable.

Try answering the following questionnaire :—

1 Do you keep more than thirty hives ?
2 Do you manage more than one apiary ?
3 Do you sell equipment at a profit ?
4 Do you use business notepaper ?
5 Have you a job in which bee-keeping plays a part ? e.g. a grocer's shop, a fruit farm, a tea garden, or an appliance shop.
6 Do you advertise ?
7 Is keeping bees the only job you have ?
8 Do you only keep bees at weekends and holidays, or do they occupy a considerable portion of your working week, i.e. do bees and something else constitute two part time occupations ?
9 Do profits from bee-keeping contribute a substantial proportion of your income ?

I do not suggest that a " wrong " answer to any question means the profits will be taxed. The question is just — " in your own personal and particular circumstances, is bee-keeping a business or is it a hobby." The answer may be that it depends upon your attitude of mind to the craft, and the above questions tend to help to determine it or interpret it to the Revenue.

"It depends on what you mean by 'pleasant hobby'!"

There is another side to the picture. If you have two
sources of income, the incomes are added together and you
are taxed on the total. If one is a loss, it is (normally)
subtracted from the other income. So, if the Revenue
decide to tax you on a good year, they may be expected to
play fair in a bad year and allow a reduction for a loss, and
quite a serious loss can be expected to follow an outbreak of
foul brood.

Those who are worried or doubtful might read Sect. 31 (1)
of the Finance Act 1948 (or better still take proper professional advice). Here are some extracts from the Section :—

(a)   all farming in the United Kingdom shall be treated
as the carrying on of a trade or, as the case may be, of part
of a trade . . . . . . ;

(b)   the occupation of land in the United Kingdom for
any purpose other than farming shall, if the land is managed
on a commercial basis and with a view to the realisation of
profits, be treated as the carrying on of a trade or, as the
case may be, of part of a trade, and the profits or gains
thereof charged to tax under Clause 1 of Schedule D accord-
ingly ;

(c)   . . . . .

Now, a word of warning. Every year we are all supposed
to fill in a *Return of* Income, and on it we have to put all
our taxable income. Now, if, in Mr. Jones' case, whether he
knows it or not, bee-keeping profits *are* taxable and he did
not declare them, he may find himself in very serious
trouble. If Mr. Jones can show reasons for believing his
profits were not taxable, he will probably be let off lightly,
which is to say either the profits will be ascertained or, if
not ascertainable, will be estimated, and all tax due for the
last number of years will have to be paid with interest. If
Mr. Jones can show no good reasons for non-disclosure, as
it is called, there are a mass of perfectly horrible penalties
laid down, and, if it happens that Mr. Jones is already " in
trouble " with the Revenue for some other non-disclosure,
he may expect some of the more horrible penalties. These
penalties, by the way, are not just airy words in an Act of
Parliament, but really are enforceable and enforced.

The fact remains, however, that the vast majority of weekend bee-keepers are not liable to tax on their profits (if any)

but if there is any doubt, it is better to be safe than sorry. All you have to do is to write to your Inspector of Taxes (quote the number on your P.A.Y.E. card), and tell him that you are keeping bees. Do not write a massive great letter, as he is a busy man, but just say that you keep so many, that you keep them at weekends only if such is the case, and that, as you regard them as a hobby, you would like to be informed if the Inspector is prepared to agree with you. If you get a reply agreeing, remember that it is only regarded by him as a hobby as long as you stick to the numbers, etc., you stated.

The Capital Required

Prices charge upwards so fast these days that no sooner are estimates given than they have to be revised. I have a catalogue in front of me, showing a National Hive complete with two supers and frames fitted with foundation at £8 retail, so if the price is different next year, the figures can be adjusted.

		£
Twenty Hives with frames and foundation		
Double Nationals with 3 supers ... £160		
Modified Dadant with 2½ supers ... £140		
16 x 10 Hive with 2½ supers ... £130		
average (say)		145
Twenty Nuclei on 4 frames, less the cost of frames		
included above		60
Extractor		20
Bottling tanks, knives, uncapping tray, pails, etc.		8
Feeders, seven large and a few small... ...		5
Smokers, veils, escapes and small equipment ...		5
Stands		5
Queen Rearing Set-up, complete (say) ...		12
		£260

To this must be added the shed and trailer if you have either or both.

This is an imposing total for a hobby. It is possible to start with less, but sooner or later, this or more will be the total, or else you will have to spend quite a lot more time bee-keeping ; for example, a little extractor and bottling tank means several more days extracting, and fewer supers means extracting more frequently.

Discounts

Appliance makers, in fact nearly all manufacturers, are prepared to grant discounts to certain buyers. If you write and order a hive and suggest you be allowed 25% off retail price, you will get nowhere. Generally speaking, you will be given a deduction off retail prices if either,

You order a large quantity at once,

You can show you are a commercial beekeeper,

You will act as an agent and sell goods to others.

Of these alternatives, the first is simplest ; you just ask one or more firms to quote for 20 hives, 1,000 frames, etc., in the flat and considerable reductions may be expected. This will not suit the man who intends to increase by six hives a year until he reaches the manageable limit. The second alternative does not apply to the weekend bee-keeper. The third alternative is thus left and it is not as bad as it sounds. Find two or three other bee-keepers similarly minded, and agree that they will order their goods through you as a co-operative arrangement (remember if you yourself make a profit this way it is taxable and so is the rest of the bee-keeping venture). Play fair with the manufacturer and tell him what you are up to ; it is only right that he should refuse trade terms if you are going to put him to a lot of trouble with small orders. Simplest of all is to ask a commercial bee-keeper to add a dozen hives to his next order. He may be kind enough to do it for a little or no profit, but do not be surprised if he will not.

It always surprises me that big honey producers waste so much time to help hobbyists, and waste really is the word. It is nothing out of the ordinary to find a commercial man lecturing, advising verbally or by post, and sometimes actively helping hobbyists, for which they get no tangible reward. I was surprised to read the views of a man who saw fit to write in the Bee Press that the big honey producers think only of their own big (?) profits. In my and my friends' experience, this is not the case ; but do not *expect* commercial men to do too much for nothing and remember what is fun for you is " shop " for them.

Profits

Better men than I have said 40 lbs. a hive is a good average crop. Perhaps it is so for the commercial man. It is certainly too high for the inexperienced. I think it is on the low side provided you are not keeping just a few more hives than can properly be attended to. However, let us take 40 lbs. as a basis. The price of honey is another tricky point, and I propose to get out of this difficulty by taking 2/4 average for the honey—that is the honey not including the pot. This is not the figure I work on, but what I regard as reasonable for a man with 20 hives to be able to get if he tries. It may be too high, but 40 lbs. for that matter may be too low. At this point, I must make it clear that these figures are not related to the profits of commercial undertakings. The next assumption is that a hive lasts 10 years. Parts of it last more and floorboards which have staples driven into them usually last less. Lastly, I am leaving out hiring fees for pollination. So, the Profit and Loss Account for 20 hives " may " look like this :

	£		£
To Sugar ...	8	By Honey Sales	
,, Hired help ...	5	@ 40 lbs. per hive	
,, Sundries and Rent	7	@ 2/4 per lb. ...	94
,, Depreciation ...	20		
	£40		
,, Profit ...	54		
	£94		£94

The depreciation figure @ 10% of the total assets is to cover the average cost of renewals of hives and equipment which will be light while everything is new, and heavier later.

This works out at something over £1 per week for a hard Saturday afternoon's " work " @ 6/- an hour.

You may get fees for pollination services. You may sell queens or bees. You may be able to get a crop of heather honey and find the time to deal with it. In fact, one way or another you may be able to increase the profit, but it is most unwise to bank on paying back the capital outlay out of less than five years' profits.

I will give two more examples, for 25 and 30 hives :—

	25 Hives	30 Hives		25 Hives	30 Hives
	£	£		£	£
To Sugar ...	10	12	By Sales	117	140
,, Hired help ...	8	8			
,, Sundries and Rent ...	8	8			
,, Depreciation ...	24	27			
	£50	£55			
,, Profit	67	85			
	£117	£140		£117	£140

That is a far better picture.

Add another £1 or £2 per hive for pollination fees less expenses, and another £2 per hive for heather honey, make the price per lb. up to 2/6 and the " average " net profit for 30 hives exceeds £200 a year, and then we leave the world of realities altogether :—

Profit and Loss Account for 30 hives.

	£			£
To Sugar ...	12	By Sales :		
,, Hired help ...	12	Nuclei	...	40
,, Sundries ...	9	Ordinary honey	...	150
,, Depreciation ...	27	Heather honey (less carting)	...	60
	£60	,, Pollination (less carting)	...	60
,, Profit ...	250			
	£310			£310

It cannot be done and this is why. In the first place, it represents nearly 1,800 lbs. of honey and if you get that amount, you cannot hope to sell a very high proportion retail, and that means less receipts. Secondly, if you are going to manage 30 hives at weekends, unless you are abnormally clever and never play tennis or be otherwise sociable, you will get less honey per hive. Thirdly, you will find there is scarcely enough time to extract clover honey from 20 hives and the idea of adding a further crop of heather honey with all the abominable work attached to that stuff is quite out of the question unless you are running it as a small commercial concern and have the time available other than at weekends. Fourthly, the sale of nuclei without reduction of crop presupposes Autumn increases and wintering in boxes ready for spring delivery, a profitable plan, but one which adds quite a lot to the already overburdened time for Autumn management of what has now become forty hives.

There is no doubt at all, and experienced bee-keepers will, I am sure, agree, that for many reasons the above figures are nonsense. The kintergarten problem : — " If you can make £5 profit from one hive, how much can you make from 500 hives ? Answer £2,500." — is such utter rubbish except as an exercise in mental arithmetic that it is past comprehension that anyone could believe such things, and yet equally ridiculous assertions have been promulgated before. However, although I may not be supported by all commercial bee-keepers, I do maintain that a weekend bee-keeper with a score of hives, *provided* he goes about his hobby in a practical, purposeful and efficient manner, can make quite an acceptable addition to his income, and the income from bees will go on after retiring from business.

Financial Records

You may well say, " Here's an Accountant telling me to keep books for my hobby—what nonsense." I quite agree, if you do not want to; do not do so, but the suggestion is not nonsense.

All that has to be done is to put down all the expenses as they are made and keep a record of the sales, if you like under headings such as " in tins," " in jars," " retail," " wholesale," etc., with the weights. At the end of each year add up the expenses under the different headings and put down the totals as I have done in the examples (with any further detail you like) except that where I have put " depreciation," you will put " replacements of hives and hive parts." Do not include new hives bought to increase

your numbers, and enter as a credit a value for additional
increase in bees permanently retained. You will have to wait
a bit to write in the sales figures as the season's crop may
not have been sold. That is all an amateur need bother about.
He need not even do that, but, and here I speak with some
experience in these matters, hardly anyone who does not keep
the simple accounts mentioned above has the faintest idea
what his " hobby " costs him, or what financial gain he
makes. The cost of new hives for permanent increase and
the original costs should, by the way, go on a separate piece
of paper. It is simple and elementary for me, but I
cannot believe it is as hard for a normal intelligent being as
some of them seem to make out.

Example :

	£	s.	d.		£	s.	d.
To Stocks 1st Jan.				By Honey sales	121	3	9
19 @ £3 ...	57	0	0	,, Nuclei and			
,, Sugar ...	8	10	0	queens sold	13	7	6
,, Lorry hire ...	3	0	0	,, Pollination	10	0	0
,, Jars and tins ...	20	4	6	,, Stocks 31st			
,, Wages ...	5	2	0	December			
,, Sundry renewals	8	6	9	22 @ £3 ...	66	0	0
(less wax)							
,, Rent ...	1	5	0				
	£103	8	3				
,, Profit ...	107	3	0				
	£210	11	3		£210	11	3

This assumes a fairly good year. The item " renewals "
includes foundation and it is assumed the wax is credited
against goods. An increase from 19 to 22 hives is shown. If
the number of hives on the 1st January was 22 and at the
end of the year it was 19, less profit would normally be
shown. If this was due to winter losses, a reduction in profit
should be expected. If the reduction was caused by uniting
or " skyscraping " colonies, the loss in value of stocks
should be offset by a greater sales figure.

There is a means of completing the year's accounts without waiting for the crop to be sold. Substitute for " Honey Sales " the description " Honey produced @ 2/6 " and put in the figure as soon as extracting is over. When the crop is sold you can see whether you did better or worse than 2/6 or whatever estimate you made. In technical terms, the 2/6 would be termed as " Standard " and the amount by which the actual sales differ from the standard is termed " Price Variance."

Hive Records

This also comes under the general heading of " paperwork."

When a bee-keeper approaches a hive he wants to know the " case history " or what he expects to find when he opens it, and, if you work it out, these two things mean precisely the same thing.

If you are a most experienced bee-keeper, you will open a hive, diagnose the condition and after a moment's thought prescribe a treatment most suitable in the circumstances. The ordinary weekend bee-keeper opens the hive, goes carefully through it perhaps twice, and perhaps has a second glance at the record and then has to make a decision as to the most suitable treatment. Most of us have often wanted a few minutes grace to chew the problem over, but with a hive open and rapidly filling with foragers, this is " off." So a decision is taken and action upon it started and possibly finished and possibly a different decision taken when half way through. Anyway, the hive gets put together again and in the case of at least one hive a week, most of us wish we had done something else with it.

I have a system which goes a long way to help with this particular trouble. My record book is a foolscap sized book with a page for each hive. It shows the number at the top and the queen's pedigree (of which more later) and a note about each examination. During each week, I take a double sheet of foolscap with two lines for each hive and pin it on a drawing board. It gets " written up " one evening during the week or even during lunch or some odd moment. In the first line I write what is " expected " or just a note if all was well last time, followed by proposed action if what is

When a bee-keeper approaches a hive, he wants to know
what he expects to find when he opens it!

expected is actually found or if something should be done in any case. The second line I fill in, in pencil, with what is done and that gets put into the Record Book later, when the next weekend's "job-sheet" is prepared. I put the order of examination which is a help in more ways than one, e.g. if rain comes on half way through it is a relief to know that the hives which are expected to be trouble-free are the only ones not examined.

Example :

Exam. Order	Hive No.	
5	1	Cells with eggs last week were destroyed. If rebuilt, move hive to divert fliers. *Swarming apparently given up.*
4	2	O.K. last week. See if fit to receive fliers from No. 1, if required. *O.K. Needs super soon.*
14	3	Fliers from 15 added last week, also super. *Very strong and doing well.*
8	4	Mated queen introduced last week, see O.K., if not, unite with No. 2, unless No. 2 has No. 1's fliers, in which case unite with No. 11. *Queen accepted and laying.*
2	5	Queen cells with grubs last week, destroyed. Removed sealed brood and gave room. Must not be moved as adjacent hive not fit to receive fliers. If swarmed remove all sealed cells ready for introduction mated queen next week. *Presumed queenless. Sealed cells destroyed.*

I do not mean to suggest that such a performance is ever expected from my own hives 1—5 ! Nor do I suggest that the proposed treatments are best. Here the examination order helps, because obviously No. 2 hive wants looking at before No. 1 or the bee-keeper will not be sure what he *can* do to No. 1. The note against No. 5 saves hunting up the records of the other hive on the same stand.

Queen Pedigrees

Each queen has a code number which is noted in the record of the hive she is in. The method is not original and is as follows. Q.6bg/yy/cl/50l. There was once a breeder queen No. 6 and her second daughter was Q.6b which was also used as a breeder. The queen in question is Q.6bg which is the 7th daughter (g) of Q.6b. " yy " means she is the normal Italian yellow, while a single y means rather on the dark side and a dash (-) indicates a dark throwback. Cl means she is clipped, and 50l means she was mated (late) in 1950 after the main laying season. Q.6bga/y/-/51 would be the first daughter of the above queen which is a little on the dark side and has not yet been clipped. If you get a number like Q6bgadmm/etc., and she is to be used as a breeder, it is easy enough to call her Q7 and start again. If (say) Q6d is not being used as a breeder, but one of her offspring is in a hive due to a virgin emerging which was not replaced, she gets booked as Q6x, which shows the line and the rest is not needed.

If you are selling queens, it is a good idea to mark each one with a tinfoil disc with a number on it and tell buyers to quote the number in correspondence. A notebook giving the parentage of the queens sold can be kept.

Mating Hive Records

There are various plans. Snelgrove has a clock with concentric dials and the " hands " show the condition. Jay Smith has little blocks of wood which show the condition according to whereabouts the blocks are placed, left or right, etc. The man with from half to a dozen nuclei need not use any fancy arrangements unless he wants to. A notebook, page for each nucleus, is all that is wanted.

Hive Record Cards may be useful here, a new card each time a queen is removed, but the disadvantage is that you cannot tell without quite a hunt round how many newly mated queens are available or likely to be so.

A useful tip is to put a queen cage on the lid of each nucleus with a mated queen known to be in it at the start of operations. Then, if you are going through a honey producing stock and require a queen, you can see where to go for one in a hurry.

~Specimen notebook entries for nuclei :—

Nuc. 1.	4	5
2/7 Virgin caged (ex Q.4ab)	Virgin caged (ex Q. 4ab)	Virgin caged (ex Q.4ab)
9/7 Accepted	Accepted	Rejected, unsealed cells left.
16/7 Laying	Not in lay	Fresh virgin Caged (ex Q.4ab)
23/7 Removed to No. 14. Cell in (Q.3b)	Laying	Accepted
30/7 Emerged	United. See Stock 10. (Q.4aba)	Laying. Removed Sold (81) Virgin caged (ex Q.4ab)
6/8 Not in lay		Accepted
13/8 Not in lay		Not in lay.
20/8 ? Q. lost		Laying. Q. to No. 4.
27/8 Queenless. Shook bees into No. 18 and closed down		Bees United to No. 18 (Q.4abh)

CHAPTER IX

POLLINATION

THE only question from a practical point of view is : " at a
given price for hire, how many hives per acre is the economic
number for each type of crop?" I think it is to be regretted
that at least one official adviser makes the recommendation
that about one hive per acre is required. Surely this is arrant
nonsense. It must depend upon the total number of pollinating insects already present, so that in districts well popu-
lated with bumble bees, honey bees, wild bees, beetles, flies,
etc. no importation of hives may be needed. Insects were
counted when visiting blossoms at Wye, Kent, by Mr. C.
H. Hooper, and these are some of the figures :

	Honey Bees	Bumble Bees	Wild Bees	Flies	Other Insects
Pear	252	19	15	58	7
Plum	55	34	36	17	1
Gooseberry	73	29	2	5	1
Black Currant	55	18	8	—	1
Cherry	106	92	16	—	1
Apple	479	54	23	27	160

One bumble bee is probably worth several honey bees when
it comes to pollination, but it is often found that in a cold and
wet spring, orchards containing hives will receive fairly
complete pollination through the efforts of a large force of
workers which issues during occasional spells of sunshine,
and completely outshines the efforts of the other insects.
Remember, fruit blossom must have sunshine for pollen
transference to be effected—even with hothouse peaches.
Hire of bees may be compared with an insurance policy ;
insurance against a bad pollination season.

In Suffolk I put a dozen hives in a black currant field in one parish, and five in another field only 5 miles distant. My efforts at counting for ten minutes were:—

		Honey Bees	Bumble Bees	Other Insects
Parish St. M.	5 Hives	110	3	2
Parish W.	12 Hives	4	6	1

Last year the owner of the first orchard doubted whether the bees did any good because whether I brought my bees or not, you could " hear " the field. The owner of the second orchard also doubted whether my bees helped because they did not appear to work the blossom. However, the assertion that one hive per acre is the right density probably does little harm. Incidentally the field with 12 hives *in it* had a very poor crop of currants.

What I think does no end of harm is this sort of thing :

YIELD 1952
Bees used for
Pollination

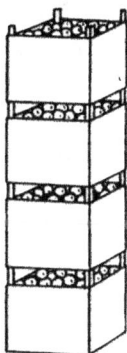

YIELD 1951
No Bees

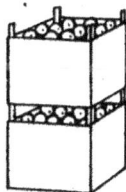

The sort of picture, which, however true, does far more harm than good.

No sooner is this sort of thing published than plenty of other fruit growers produce pictures showing precisely the opposite result. There are plenty of fruit growers who can (and will) produce honest figures to show that bees apparently have no effect on crop yields, and to give them an opportunity to do so is to do a grave disservice to bee-keeping.

Another bit of advertising, which probably does no harm, is to take a beautiful photograph of one bee burrowing into one blossom and suggest that a million bees on the same task complete the job of pollination. They do not, and there is a lot more in the foraging behaviour of bees in relation to pollination than many people think.

Bee-keepers believe bees do no end of good to apple blossom, and experiments in America seem to prove without any doubt that the honey bee is an efficient pollinator. Apparently a tree called the buckeye yields nectar poisonous to bees, so wherever these trees are indigenous, no bees are to be found. Orchards in these " buckeye districts " had poor yields of apples, but when bees were moved into the orchards on hire and removed before the buckeye blossomed, enormous increases in yields were recorded—so enormous that the value of bees could not possibly be questioned.

However, quite a number of fruit growers think that bees are not necessary, or at least that equally good results can be obtained by dabbing a rabbit's tail into the blossoms. Bush had something to say on those lines but spoilt his argument by saying that bees were difficult to keep.

For every man who makes his living from bees in this country, there must be hundreds who make their living from fruit growing. Put another way, most fruit growers are commercial men and most bee-keepers are not. If a fruit grower pays for the hire of bees he does so for one reason only—because he thinks he will make more money. When fruit growers meet to discuss business, if one man has stopped hiring bees and noticed no difference, the others will think twice before spending £50 or so each year for hire of bees.

I hired my bees to a fruit grower and I noticed he had to pay for apple thinning—the job had been over-done. That year extra bees were neither necessary nor desirable in his orchard.

I remember a fruit grower telling a bee-keeper that he got just as many apples since all his own bees died as he did when the colonies were alive. Of course, this is rank heresy, and the bee-keeper knew for sure that his leg was being pulled. After a while it began to dawn on the bee-keeper that the fruit grower did (dare to) believe that bees were no use. I felt sorry for the bee-keeper, he got the worst of

"Fruit-growers are not **entirely** convinced of the value of bees!"

the argument because he had no argument. If someone had told him that the moon would hit the earth tomorrow, he would have regarded that person as a joker or a lunatic and would have no argument beyond stating that it would not be the accepted order of things. But in this case the fruit grower was not joking and did not look lunatic, but was (and is) a calculating animal and had no use for " accepted beliefs " when it came to signing cheques.

The doubting fruit grower will say " prove to my satisfaction that bees will increase my profits and I will pay for them." The usual reply " of course they will " is no longer good enough. A bee-keeper might reply " you may hire my bees at £3 per colony and if you do not get more bushels of apples than you did last year there will be no charge." I do not agree with that idea, but at least it would prove that the bee-keeper had confidence in his wares. If you do make such an offer, first find out about apple trees of a class known as biennial bearers.

Not infrequently a grower is completely unaware that there is an apiary within flying distance of his trees. He glibly asserts that he gets along quite well without bees when he is enjoying adequate pollination services free. One such man rose at a meeting and asserted that there was not a single hive within two miles of his orchard. At once his neighbour informed him that within 200 yards of his boundary, was an apiary of 30 hives. It must also be remembered that a man who hires bees, and has them sited fairly near his neighbour's orchard, may well be providing an adequate pollination service for that neighbour as well as for himself.

I regret that in the case of seed growing also I have been unable to get any figures proving the value of bees. One local seed grower hired bees and was made to promise that he would not use poisonous sprays on his crop (of swedes). The result was that the field near the bees gave less seed than the others owing to the damage by the pollen beetle. The distant fields gave a high yield—they were sprayed.

My bees gathered a lot of red clover honey in 1950 from a field about 50 yards away. The farmer told me I ought to give him some of the honey my bees got off his property, so I replied that he ought to pay me for the extra seed which he got as a result of the thorough pollinating jobs my bees had done for him. About a quarter of a mile beyond the field my bees worked so noisily was another field of red clover. That more distant field yielded a tiny fraction *more* seed than the near field. I used to wander round the near field because it was heartening to see so many bees obviously getting a lot of nectar. There were bees in every square yard of the field, and the space between two May trees by which the bees mostly entered the field was a congested thoroughfare, while the distant field had comparatively few bees in it. I told the farmer that he must have harvested the far field better, and he said " No." I said he must have had more damage to the seed in the near field from mites. I said the first cut for hay was delayed so the seed did not have time to ripen. But it was little use reasoning ; the farmer was convinced it was fair comparison and he appeared to me to believe in his findings. I told this story to a County Instructor in Bee-keeping. His opinion was that the farmer was probably telling lies to get some free honey. In other words he thought the near field " must " have yielded more. But did it? I wonder. I suspect the far field was better clover growing land.

Surely bee-keepers should do something to prove the value of bees as pollinating agents ? We know they are valuable, but we cannot produce any facts and figures to show the most economical number of hives per acre of a given crop in a given district. Fruit and seed growers are a far greater power in the land than we are, and they are no fools. If figures are produced they are sure to do more harm than good if they cannot stand up to a careful examination ; shrewd commercial growers are not going to believe a lot of figures because they are in print.

I think it would be a good thing if bee-keepers who are trying to hire bees learned a little about fruit growing. Fruit growers I have met appreciate it if a little intelligent interest is taken in their orchards. I have had many pleasant discussions with fruit growers, and one result is that one orchard is now planted in accordance with a theory worked out between the owner and myself. It used to be recommended that an orchard of Cox (x), with Grieve (P) as the pollinators, should be planted according to the following pattern :

```
X   X   X   X   X   X   X   X   X   X   X   X
X   P   X   X   P   X   X   P   X   X   P   X
X   X   X   X   X   X   X   X   X   X   X   X
X   X   X   X   X   X   X   X   X   X   X   X
X   P   X   X   P   X   X   P   X   X   P   X
X   X   X   X   X   X   X   X   X   X   X   X
X   X   X   X   X   X   X   X   X   X   X   X
X   P   X   X   P   X   X   P   X   X   P   X
X   X   X   X   X   X   X   X   X   X   X   X
X   X   X   X   X   X   X   X   X   X   X   X
X   P   X   X   P   X   X   P   X   X   P   X
X   X   X   X   X   X   X   X   X   X   X   X
```

It can be seen that every Cox is adjacent to a pollinator.
(Fruit growers seem to use the word "pollinator" when
referring either to the insect or to the trees planted especially
to provide suitable pollen for self sterile varieties). I think
it is now considered that one pollinator in nine is insufficient.

Our theory is based first on the fact that an apple orchard
is an attraction to insects in the neighbourhood, so insects
will fly to the orchard from the surrounding area. Secondly
the trees on the South and East borders will be in the sun
first each day and the trees further in will be to some extent shaded until the sun has risen appreciably. It would
therefore appear to follow that honey bees, bumble bees and
other insects coming into the orchard will alight first on the
trees in the outside rows with a preference for those on the
Southern and Eastern borders. As the blossoms on these
trees become congested and sucked dry, the insects should
logically migrate towards the middle of the orchard carrying
with them the pollen from the trees on the borders. *Therefore the Southern and Eastern border trees should all be
pollinators and the other border trees should have a
good proportion of pollinators.* To us this seems logical,
so it is being tried out.

Experiments seem to prove that old field bees tend to work
a small area, and one may get "fixed" on say half a bush
apple tree. However often it visits that tree, it will do no

pollinating except when it brushes against other bees in the hive and collects a few pollen grains which it may leave behind in the next blossom it visits. Younger field bees tend to leave a patch of blossom more readily than older bees if it becomes overcrowded with insects. For this reason I am inclined to the view that the hives should be sited in a group. The adjacent trees then become overcrowded, the young bees change their foraging area and in so doing they take pollen to other areas.

Fruit growers will spray fruit trees, sometimes when in blossom. If you are going to remonstrate, learn the subject or you will find the fruit grower will say something like " if I do not spray now, the sawfly I have seen will do damage which I expect will cost me £1,000 or more. It is pointless to pollinate apples for bugs to eat, and which are worth most —your bees or my apples?" I was warned by telephone that nicotine was to be used on open apple blossom the next day and I removed half the hives. A week later the Norfolk County Instructor (F. A. Richards) was unable to find any difference between those which had been " nicotined " and those which had not. If anything, the ones which had stayed on and braved the spray were better because they had gathered more apple honey. Dispersible sulphur appears to have no effect, and lime sulphur I do not worry about. Lead arsenate is obviously the spray to be frightened of, but I do not think many growers use it on open blossoms. I feel sure it is better not to get steamed up, but to ask the fruit grower as a favour if he would spray as early in the day as possible. Pull up some grass and throw a pile in front of each hive. By the time the bees have worked free the spraying will be over and the most dangerous time passed.

While writing of apple blossom, I am taking the opportunity to mention dandelions which interfere with the pollination of apples to an extent out of all proportion to their numbers for four good reasons.

1 The dandelion comes first and old bees tend to get " fixed " on them.

2 The orchard may be sprayed before the apple blossom is out. The dandelion then becomes poisonous to bees.

3 There is more nectar per dandelion. There is more pollen per dandelion. The nectar is stronger. The pollen is

Bees and Other Insects that Visit Fruit Blossoms.

(Photographs by Fred. Edenden, of Wye, Kent)

Plate I.

1. Hive Bee ; Bumble Bee Queens :—2. Large Earth (Bombus terrestris) ; **3. Stone** (B. lapidarius) ; **4. Early-nesting** (B. pratorum) ; **5. ditto, worker ; 6. Small Garden** (B. hortorum) ; **7. Small Earth** (B. lucorum) ; **8. Common Carder Bee** (B. agrorum) ; **9. Red-shanked Carder Bee** (B. derhamellus) ; **10. Shrill Carder Bee** (B. sylvarum) ; **11. Large Carder** (B. muscorum) ; **12. Large Garden** (B. ruderatus). **Cuckoo-Bumble Bees :—13.** (Psithyrus barbutellus). **Smaller Wild Bees :—14.** Osmia rufa ; **15.** Andrena albicans ; **16.** A. fulva ; **17.** A. gwynana ; **18.** A. nigroaenea ; **19.** A. thoracica ; **20.** A. dorsata ; **21.** A. Nana ; **22.** Halictus cylindricus ; **23.** H. rubicundus ; **24.** H. morio ; **25.** Nomada ruficornis ; **26.** Specodes gibbus. **27. Solitarp Wasp** (Odynerus callosus). **Queen Wasps :—28.** Common (Vespa vulgaris) ; **29. German** (V. germanica) ; **30. Wood** (V. sylvestris). **Ants :—Black Garden Ant** (Lasius niger) **31. male ; 32. female. Saw-flies :—33. Apple Sawfly** (Hoplocampa testudinea) ; **34. Gooseberry Sawfly** (Nematus ribesii) ; **35.** Raspberry Sawfly (Macrophya rustica) ; **36. Moth** (Anarta myrtilli).

Plate II.

1. Fungus Gnats, Sciara ; **2. St. Mark's Fly** (Bibio Marci) ; **3. Fever Fly** (Dilophus febrilis, L.) ; **4. Fever Fly** (Bibio hortulanus) ; **5. Bee Fly** (Bombylius major) ; **6. Dance Fly** (Empis opaca) ; **7. Dance Fly** (E. livida) ; **8. Robber Fly** (E. tessellata) ; **9. The smallest of the Syrphid Flies** (Platychirus albimanus) ; **10. Syritta pipiens ; 11. Hover Fly** (Syrphus ribesii) ; **12.** Rhingia rostrata ; **13.** Heliophilus floreus ; **14.** Volucella bombylans ; **15.** Pellucid Volucella (Vollucella pellucens) ; **16. Drone Flies :—17.** Eristalis pertinax ; **18.** E. tenax ; **19.** E. sepulcralis ; **20.** E. nemorum ; **21.** E. intricarius **(larvæ rat-tailed maggots) ; 22. Narcissus Fly** (Merodon equestris) ; **23.** Myopa polystigma ; **24. Yellow Cow-Dung Fly** (Scatophaga stercoraria) ; **25, 26, 27. Anthomyiid Flies ; 28. Blue Bottle** (Calliophara erithrocephala) ; **29. Blow Fly** (C. vomitoria) ; **30. Green Bottle Fly** (Lucilia cæsar) ; **31. House Fly** (Musca autumnalis) ; **32. Common House Fly** (M. domestica) ; **33. Flesh Fly** (Sarcophaga carnaria). **Beetles :— 1. Raspberry-beetle ; 2, 3. Two-spotted and Seven-spotted Lady Bird Beetles ; 4. Blossom beetle ; 5.** Micrambi vini ; **6.** Aphodius inquinatus ; **7. Click Beetle ; 8. Apple Blossom Weevil ; 9.** Rhynchitis aequatus ; **10. Phyllobius oblongus ; 11. P. pyri ; 12. Earwig ; 13. Small Cabbage White Butterfly.**

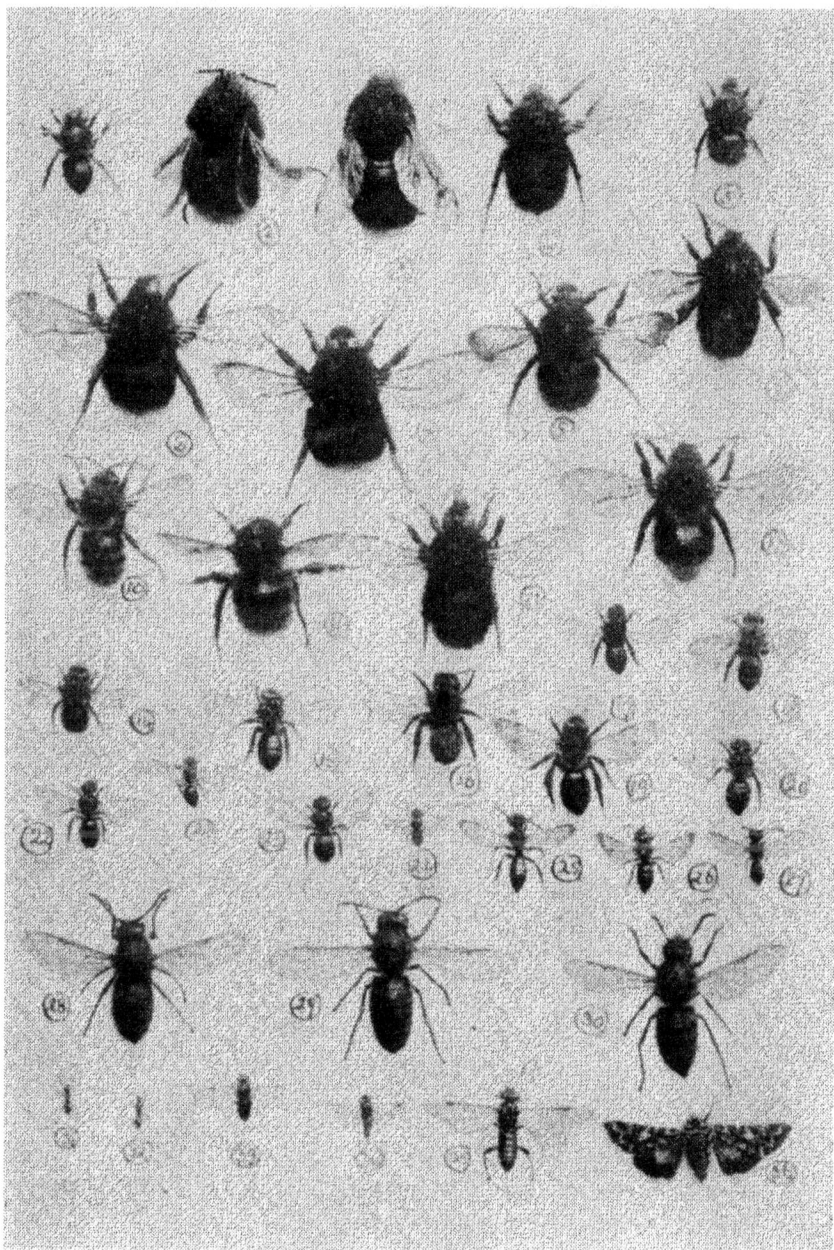

PLATE ONE

PLATE TWO

more easily collected. In fact the dandelion is generally more satisfactory for bees.

4 The dandelion, being near the ground, can be visited more easily on windy days.

Fruit growers do not like dandelions, and if you tell them all this, they will be pleased to wage a more intensive war on them (mainly to your disadvantage perhaps, but you are not paid to have your bees visit dandelions).

It seems to me that one of the reasons why the value of bees as pollinators has not been tested is for financial reasons. Fruit growers would have to send in special reports of crop yields, type of apple, numbers of pollinator trees, number of hives, densities of pollinating insects, etc., or else someone would have to collect data from a number of farms suitable for comparison, lending bees in some cases, to ascertain comparative results. But who is to pay? Such a survey might cost a pound per farm or more, and statistics of this sort would not be of much use unless there were about 500 farms involved spread over not more than two counties, and that for fruit alone.

Something might be done on a small scale, and I have some suggestions to make as to how it might be done. If the annual yield per acre of two separate orchards, both without hives, was known over a period of years, and then one of these orchards only had bees introduced, the results could be recorded, as shown in Graph "A"

Graph "A"

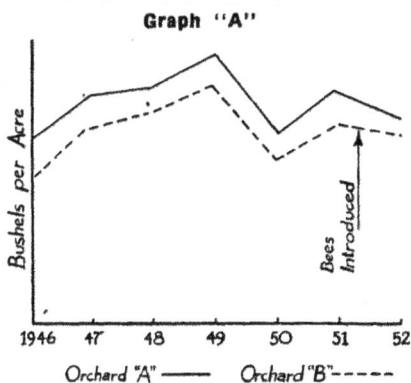

Orchard "A" ———   Orchard "B" - - - - -

In this imaginary example I have made the yield drop after introducing bees into Orchard "B" but it is plain that the yield has increased in comparison with Orchard "A".

By this method some of the factors which cause the annual fluctuations in yields would be eliminated, but not by any means all of them. Orchard "A" for instance, might have had a reduction in yield due to red spider. These other fluctuations could be ironed out by plotting the figures of the yields of more orchards.

If the yields of about a dozen orchards are known, and then half of them introduced bees, the graph would be considerably more accurate. The selected fields should have the same variety of apple and be as similar as possible, i.e. about the same number of pollinator trees, and trees of about the same age. Biennial bearers are unsuitable. As the yields vary so much it would be better to plot the differences between the groups only, otherwise the graph would be too large to be clear.

Graph " B "

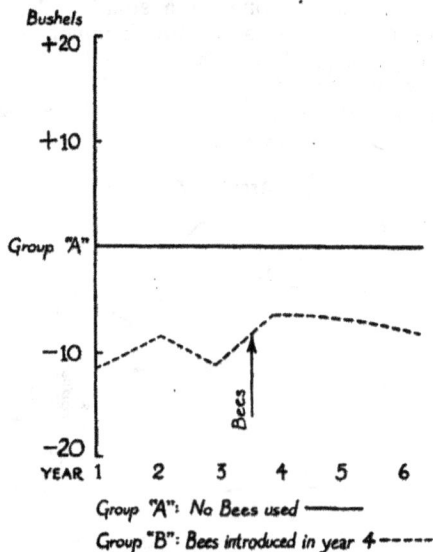

Group "A": No Bees used ———
Group "B": Bees introduced in year 4 - - - - -

In Graph "B" the average results of six orchards with no bees are compared with the average results of six orchards which had no bees up to year 4 but where they were used in years 4, 5 and 6. I have made the results show no appreciable difference between the two groups, and this would indicate that in that district only, bees were of little use for those particular years.

It should be noted that an extra yield of only five bushels per acre would pay for the hire of a hive, and anything over five bushels would be clear profit to the grower.

An experiment on this scale would obviously be expensive and entail a lot of work, but an experiment on a smaller scale, although not so accurate, should be informative. Suppose there were two orchards, F and G, growing the same variety of currants, F only having bees on it, and that the severe winter of 1947-48 killed most of the bees the results might well show that for several years F averaged the same yield per acre as G, but that in 1948 F, now without bees, yielded "x" lbs. less currants than G. If the net value of "x" lbs. currants exceeds £3 (the cost of hiring one hive) it would seem that bees are worth hiring.

Thus any grower who has either started using bees or has ceased to use them, can prepare these figures, provided he knows the yield per acre of his own orchard and that of another grower in the same district for the same years.

I wish I had more constructive suggestions to make. Something should be done to prove the value of bees as pollinating agents in this country, but I cannot pretend to know the best way of going about it.

CHAPTER X

ODDS and · ENDS

The Library

THE Reviews in the Bee Press should be watched, and you will then have some idea which new books should be taken with a grain of salt. Books can usually be borrowed from the Association Secretary, or, free of charge from County Library. The County Library is financed from the Rates, so in a sense you have to pay for the books whether you have them or not. If the book you want is held in the County there is no charge for borrowing it, but if it has to be got from outside the County, then postage must be paid on it. I believe a good plan is for a few bee-keepers to get together and demand all new books on bee-keeping, each on separate demand cards. This probably ensures a copy being kept '' in the County.''

As I have said before, read all books. You will soon see which you want to buy. However good the library service may be, it is really necessary to have a few books in the house. The following list of books is not intended to be complete in any way.

Honey Farming (Manley). Tells you how to manage a chain of apiaries. Follow the '' drill '' laid down, regarding your concern as equivalent to one of his out-apiaries, and you will not go far wrong.

Manual of Bee-keeping (Wedmore). *Bee-keepers Guide* (Digges). These two are taken together. Practically every obscure or commonplace item you may ever want is sure to be mentioned.

Bee-keeping in Britain (Manley). To my mind, not quite as important to have as *Honey Farming*, but better for the less experienced.

Bee-keepers Folly (Ratcliffe). For those who are starting or who are going to expand, this is especially good. Those who are using double brood chamber hives need a copy.

The Bee Craftsman (Wadey). It would be a pity not to have this little book. It does not cost much, and is full of interesting and sound theories.

Swarming, etc. (Snelgrove). A most useful little book, whether you intend to follow the system, or whether you do not.

Introduction of Queen Bees (Snelgrove). Useful and thought provoking.

A.B.C. (Roots). *The Hive and the Honeybee* (Grout). Do not omit these two from the list of books to be read just because American conditions are different from ours. No end of a lot can be learned from these books.

Art of Bee-keeping (Hamilton). *Successful Bee-keeping* (Wedmore). Good elementary guides.

Queen Rearing. Books by Abbott, Laidlaw and Eckhert, Jay Smith, and Snelgrove should be read if queen rearing is to be undertaken.

Examinations

The B.B.K.A. examinations for proficiency in apiculture include questions on any aspect of bee-keeping and bees. Some years ago, I decided to take these examinations. The paper on bee-keeping I had could only be described as dead easy. I did not get full marks ; if my memory serves me right, I lost three marks somewhere. The second paper contained a number of questions about anatomy, in one of which I was expected to draw parts of bees' insides, and label bits of them in Latin. There was a request for a picture of some flower, labelling all the parts, and I was also expected to know " who invented what " some fifty years ago. After that, I gave up thinking about Examinations.

There is a qualfication called " Practical Bee-keeper " you can go in for. I have not. It is described as suitable for those who are not cut out for written exams., so naturally I would not like it thought that such a description applied to me.

Let it not be thought that I wish to discourage anyone from taking these examinations. My opinions on this subject are :—

(i) Do not enter unless you really know anatomy.

(ii) Do not accept the teachings of all Master Bee-keepers. By no means all of them know much about bee-keeping, although they are all very good indeed on knowing what goes on inside bees' insides.

Heather

The nearest heather to my house is at a well-known place called Walberswick. It is on low-lying damp land, nowhere near any granite, splashed with salt spray and, of course, it yields no nectar—it just couldn't, everything that could be wrong is wrong. However, a bee-keeper came to live near here and nobody told him about this, so when he saw the heather, he decided to take his bees there. That was in 1948 when it rained all the time and queens stopped laying in July owing to shortage of stores. By August, we were all wondering whether an extra ration of sugar could possibly do any good. But the hives at that awful heather were almost too heavy for two men to lift. Yes, it was heather honey. And nobody can understand how it could come to pass.

I tell this story, which is true, because, although I have little personal experience of heather, it seems that a lot more nonsense is written on this subject than on most bee-keeping subjects. I think this must be because as heather receives less attention from commercial men, a lot of the abracadabra surrounding that sphere of bee-keeping has not had its proper share of de-bunking. It is time it had.

Stray Swarms

Leave these alone is the usual advice. Another line is taken by some strong minded folk ; take the swarm and destroy it, thus making sure it does not spread any disease to any neighbouring apiary or hollow tree. All very sound. If you find half a crown on the king's highway, leave it alone ; you may catch leprosy. Also sound.

For the 95% of us who will collect a swarm if one arrives at some convenient place, it is as well to mention some precautions. First, hive it on foundation so the bees cannot put the honey they carry into cells. Secondly, feed syrup with sulpha in it. Thirdly, take a sample and test for acarine and nosema. Fourthly, remove the queen as she runs in and substitute a virgin ; if no virgin is available, one can be introduced next day, and if still not available, the old queen can be removed next day (before she starts to lay) and a new virgin introduced two or three days later. The object being to try to get a break in laying. Another idea is to put

the swarm into a hive with foundation in all frames but one,
and that one a rubbishy old comb waiting to be melted. All
honey the swarm carries will be put in that frame that night,
and the next day it can be removed (do not shake it) and put
into the solar extractor at once.

Repairing Damaged Combs

If you have no end of time to spare, you can repair them
by cutting out the dud portions and fitting in fresh foundation cut to fit. Far quicker and simpler, is to cut out the
bad parts and put the frame in the middle of a super between
two capped frames where the bees will fill it out with worker
cells. Quicker still, is to put the frame in the solar extractor
and fit fresh foundation. Frames containing too much drone
comb can be collected together and put into the hive from
which you would like most drones to be reared, but, of
course, there is a limit to this.

Emptying Combs

Before rendering combs, it is as well to get the honey out.
Another occasion is when you have brood frames full of honey
and no extractor to take that size. This operation is very
much harder than is usually realized. The recommended
advice to put the frames to be emptied above the crownboard, uncapped, spaced wide, higgledy-piggledy, etc., very
seldom works. If it does not work, try soaking a few combs
at a time in water and then put them over two crownboards
upside down. Then feed the water ! In times of summer
dearth, the bees will clear frames with little trouble, but in
the normal summer time, I find it just does not happen, and
just before extracting time the bees always refuse to empty
combs probably out of sheer cussedness. Wadey recommends filling the combs with water from a syringe, a method
which shortens clearing extracted combs to a quarter of the
normal time. I intend to try out this method.

Crops and Pasturage

It is interesting to know what plants yield in what conditions and approximately when. I used to record all flower-
ing dates meticulously until someone told me that my careful

records must get me no end of a lot more honey ! Now my records are not so neat, but I do think some advance knowledge is useful.

Of the early useful plants, I find sallows and blackthorn, especially the latter, blossom when they feel like it, and follow no set pattern. In April, I keep an eye on the gooseberries, and once these flowers appear, I know the willow catkins will start to open almost exactly a week later. The willows may begin any time between 10th April and 1st May, but if you have a gooseberry bush, you will normally get a week's notice. For me, the willow is important as it provides the first pollen and nectar of the year for early brood except the little bit from minor sources. The early apples begin near enough ten days after the willows begin and the Cox's Orange Pippin, five days later. Currants come into blossom between the willow and the apple, and so does the pear, but I regard pears and plums as being of little use in this district; there are no cherries. I have not got a list of earliest and latest opening and finishing dates for all nectar sources, and this is what I use :

D day	Gooseberries begin
D + 7	Willows begin
D + 11	Currants begin
D + 17	Early apples begin
D + 22	Cox Orange Pippins begin

In this district the first field beans come into flower fairly punctually around 20th May, and, contrary to a lot of writings, are one of the most reliable local sources of nectar. They last a month. The only other good source of nectar hereabouts, is the red clover which yields a moderate amount about every other year, and heavily once every fourth year. 1949 and 1950 showed red clover yields, and 1947 was quite exceptional. Knapweed usually yields some nectar, and white clover occasionally. Sometimes a little apple honey is stored. Bees visit the hawthorn which is a lovely sight, but I do not know if they get much honey from it here. I wish someone would grow sainfoin.

Moving Stocks

Terrible tragedies may occur when hives are transported from one place to another. Whole stocks can be lost, combs

collapse through heat, and honey may run out of the
entrance. I often hear and read of these happenings, but
nothing like that has ever happened to me. The worst
disaster was when a double-National came apart and I had
a cheap veil, no gloves, and half a pint of bees up each
sleeve. The second worst was when a roof came off and my
veil got torn while carrying the hive.

Moving bees long distances as a regular annual occurrence
is not really a job for a weekend bee-keeper at all. If you
move house, arrangements to move the apiary must, of
course, be made, and it is best to consult a commercial man
and hire the tackle.

If you are going to go a matter of five to twenty miles to
orchards or to heather, there is no need to make a lot of fuss
about it ; but if you are going to do it as simply as you can,
you must take certain precautions. First, move at night.
Choose a cool night if you can, put on the entrance closers
when the bees are all in, and start moving when it is nearly
dark. Secondly, be sure to give plenty of room. Transporting
a single chamber National hive with one crate of sections
and a strong colony in it is just asking for trouble. It is
better to travel with an empty super—that is, a super without
any frames or sections in it—and put in the frames when
you get there.

Fasten the parts of the hive together fairly securely. I
once thought propolis was good enough, but events proved
otherwise. I use four crate staples for the floor and three
or four to fasten the supers to the broodchamber. Another
method is to use slats of wood screwed into place down the
whole side of the hive, but the only time I tried this I found
it to be more trouble than staples. I wish someone would
invent an efficient, quickly appliable, simple and *cheap*
device to keep the hive together ; I am trying out a new
device this year, but it is neither simple nor cheap. Staples
are, however, quite good enough for short trips.

I close the entrances with perforated zinc by the method
so admirably illustrated in *Honey Farming*. In 1950, it was
a terrible night when I took the bees to the orchard, and
sleet was falling, so I used no entrance closers. Very few
bees came out and these only crawled a few inches up the
front of the hives.

I use no top screens, and the reason is lack of time. If moving is done at night, it means another visit next day to replace the crownboards, and I will be in my office then. As no stocks have yet been lost and no signs have been noticeable in front of moved stocks that they have travelled badly, I conclude that if no light gets into the hives, very little air is needed for a short trip.

Stings

Some people always seem to get stung. I have seen two bee-keepers near the same hive, one surrounded by furious bees and the other receiving little or no attention. The classical authors say " Drinke ye only of the beste beer " and later writers advise keeping away from bees if you suffer from halitosis. There is a product called Amplex made by Ashe Laboratories, Leatherhead, and it can be got from most chemists. It contains chlorophyll. All smell of sweat, etc. disappears (or rather never appears) after eating a tablet of the stuff **as far as the human nose can tell.** Whether the bees can still smell anything or not I do not know, but I have good reason to think they do not, and from the experiments of Von Frisch it appears that, in most cases, if human beings can detect no smell, then bees cannot either. So I would recommend bee-keepers, especially those who are inclined to be stung more than usual, to give Amplex a trial.

CHAPTER XI

CONCLUSION

MANY people say they like to keep bees for a hobby and, therefore, does it matter how they do it and how much of a mess, muddle and shambles results, as surely they are allowed to do what they like behind their own four fences ? I think it does matter ; I think these men do a lot of harm to good bee-keepers in marketing honey, in spreading disease, and in perpetrating an atmosphere of magic ; I think there are a great many too many of them, and I support the Bee-keepers Associations in their efforts, by encouragement and example, to improve the standard among small bee-keepers which is, on the whole and by any standard, not very bright.

On the other hand many I know take a pride in the appearance of their apiaries. I cannot understand why so many people who keep bees purely as a hobby do not keep them better ; all the stamp collections I have seen are as neatly kept as the owner can make them, so why should there be so much neglect in bee-keeping ?

If bee-keeping is conducted on efficient and up-to-date lines, it can be just as pleasant a hobby as any other. I am firmly of the opinion that it can be " made to pay " rather better than most of the usual run of hobbies advertised as profitable. There is no need to make it pay of course, and I expect there are many who will say that if bee-keeping is to be a " pure " hobby, any money coming in should be used to improve the apiary and generally make the hobby more pleasant. A lot of enjoyment can be had experimenting with different equipment and bees, supporting honey shows and other Association's activities. This book has been written mainly for those who like to see some profit and who do not regard making it as incompatible with the idea of a hobby.

I will conclude this book with a summary of the main points I have made.

The Hive

I find Modified Dadants to be the best hives ; not the best in every way, but the best on balance. I advise those beginning to start with two or three of those hives which have M.D. floors and roofs and have B.S. frames " warm way," and plan to have M.D. hives for further increase. Those who already have a number of National hives, are best advised to stick to Nationals unless the intended increase exceeds a dozen and also the increase is to be treble the number of hives already owned. For example, if you have six hives and mean to have twenty, change to M.D. If you have fifteen hives and mean to have thirty, it is normally better to stick to Nationals, unless you can arrange to sell some Nationals at a fair price. If you already possess an expensive extractor for B.S. frames and cannot sell it well, it is probably best to super the M.D. broodchambers with supers made to take B.S. frames running across the M.D. brood frames.

Ancillary Equipment

If you already have a moderate amount of equipment, there is no need to spend a lot of money replacing it at once. As it wears out, or as the need arises, get new equipment generally of the larger sizes listed.

The Apiary

Sometimes you can do something about the site. More often if you have a poor site you must lump it.

Management

There are no hard and fast rules. Do not lose sight of the fact that the profit depends upon the total honey produced and not on the amount produced by the best hive. If a hive is in such a condition that it is unlikely to produce any honey, it will do no good to write it off as a dead loss for the season if it can be made useful in some other way. If it can be left to its own devices to collect enough stores to winter on, it may be best to save sugar and trouble and leave it alone. Usually it is better to use the bees and brood towards helping other colonies to produce more honey or at least to mate queens.

The Bees

All the experience I have had shows that a poor strain of bees does not pay for the trouble of looking after them. I once got a queen of a local strain, and a friend of mine bought a number of colonies of that same strain, and neither of us succeeded in getting **any** honey from those bees. I have had bees from several local sources too, and I found that they are as much trouble to manage as any other bees and never give anything like as much honey as the bees which were bred from queens I bought from honey farmers. By this I do not mean that because a stock is acquired from a local non-commercial source it is necessarily poor, but I do say that every single time I have got bees from other than a commercial source those bees have shown no profit until requeened. I have helped quite a lot of other people with their bees, and, with a few exceptions, I condemn their strains as being third rate.

Harvest

You will learn by experience how best to arrange your extracting. I hope my remarks on this subject will be of some help to those who find each year that extracting time is just as terrible a time as it ever was, and those who find it difficult to get jars of honey free of flotsam without expensive equipment and without spending a lot of time doing the job.

Paperwork

In business the majority of people make more profit than they think they have, and most owners of businesses spend more money for private purposes than they realize until their accounts are drawn up. I know that, as I meet these facts in my daily work. I can only say I suspect people think they make more out of so-called profitable hobbies than they actually do. Apart from finances, this chapter contains notes on records. Those with what have been called " card-index memories " need not bother so much with records. I have ceased to trust my memory when dealing with bee-hives and I probably write down much more than most people. Even so I hold that to have too voluminous records is at the least a fault on the right side.

Pollination

Remember that fruit and seed growers are practical men, and although most of them regard bees as a necessary evil, quite a lot do not regard bees as necessary at all. It is important not to exaggerate the value of bees in pollination because the grower will not only be unimpressed but may well cease to listen to bee-keepers. From the bee-keeper's point of view, hiring bees for pollination is a matter of salesmanship, and good salesmanship does not include making promises that are not fulfilled, selling at a loss, or making extravagant claims as to worth. Please be careful what you say to these people.

Conclusion

The most important factor in profitable bee-keeping is the possession of a good strain of bee. Good hives and good management will be wasted unless the strain is good. Let us hope the average quality of bees in this country will improve.

www.ingramcontent.com/pod-product-compliance
Lightning Source LLC
Chambersburg PA
CBHW072158270326
41930CB00011B/2474